The CARVER Target Analysis and Vulnerability Assessment Methodology

A Practical Guide for Evaluating Security Vulnerabilities

The CARVER Target Analysis and Vulnerability Assessment Methodology

A Practical Guide for Evaluating Security Vulnerabilities

Leo Labaj

and

Luke Bencie

SMI

Security Management International/SMI Media

Vienna, Virginia

The CARVER Target Analysis and Vulnerability Assessment Methodology
A Practical Guide for Evaluating Security Vulnerabilities

Published in the United States of America
ISBN 978-1-7324297-0-3

Design by Michael Hentges
Cover design by Anu Narasimhan

Printed in the United States of America

FOR MY FATHER, Leo E. Labaj (1919-1945), who was killed in the final weeks of World War II. I did not have the opportunity to know my father or have a chance to share our lives together, but I remain eternally grateful, proud and honored that I share his genes. My father was buried at Long Island National Cemetery, New York.

Contents

Foreword

By General James Jones

"Jim, we must ensure that our nation's critical infrastructure is secure."

That is what President Barack Obama declared to me in the Oval Office just a few days into his new administration. America's 44th Commander-in-Chief was already well-aware of the urgent need to protect the major utilities, transportation hubs and telecommunications systems that constitute the infrastructure backbone of our country. As the President's National Security Advisor, I understood better than most the importance of defending our key resources. Serving as the 32nd Commandant of the U.S. Marine Corps, as well as Supreme Allied Commander of Europe, I had spent the previous four decades studying, and applying, military strategy. As any good general will tell you, if you can knock out the enemy's infrastructure, you severely handicap their ability to fight.

Because today's world is more dangerous than ever, we must remain vigilant. Terrorists, rogue regimes and cyber-criminals continue to threaten our prosperity; they regard critical infrastructure as a potential target. To ensure our safety, it is imperative that we maintain a solid defensive game plan. Conducting comprehensive – and more important, honest – threat and vulnerability assessments on our national framework is the first step to accomplish this objective. As Chinese General Sun Tzu wrote, "Don't depend on the enemy not coming; depend rather on being ready for him."

The CARVER Target Analysis and Vulnerability Assessment Methodology is one of the most effective tools that a security professional can use to identify and evaluate weaknesses in a critical infrastructure system. It is an efficient process, which produces a numeric score against assets, to determine their probability of being attacked. It allows the person conducting the assessment to examine a potential target through the eyes of an aggressor. I first learned about CARVER during my time in Vietnam, where it was used by both the intelligence and the special operations communities. It is time tested and it works.

Luke Bencie and his team at Security Management International (SMI) have led the charge in promoting the CARVER methodology throughout the private sector for years. Leo Labaj, SMI's Director of Critical Infrastructure Protection, helped develop CARVER during his lengthy career with the Central Intelligence Agency. In the 1970s and 1980s, Leo and other demolition experts at the CIA helped train both American and international special operators on how to use CARVER to identify targets for sabotage. CARVER remains an invaluable counterterrorism resource today and is a recommended assessment tool by the U.S. Department of Homeland Security.

After I officially retired from government service, I opened my own international security consulting firm, Jones Group International (JGI), to continue the fight. We have partnered with SMI on numerous projects, and we continue to use the CARVER methodology to help our clients identify where their operations are exposed. The CARVER matrix, along with the detailed Design-Basis-Threat (DBT) models, provide easy-to-understand figures, which neatly explain how best to allocate resources, often on a tight security budget.

In the pages that follow, Luke and Leo provide a blueprint for using CARVER within your own organization. After delivering their SMI CARVER Target Analysis and Vulnerability Assessment Training Course™ to those in the security industry for years, it only made sense that they complement that training with the definitive handbook on the subject. Whether you are looking to enhance your physical or cyber security, or wish to evaluate your strategic planning goals, this book will serve as a reference you can access each time you begin your assessments.

It would be naïve to think that our adversaries will not try continually to exploit our national vulnerabilities. You should consider the same for your own organization. If you cling to a strategy of, "it will never happen to us here," then you are unfortunately complicit in your own demise. You owe it to yourself, and those around you, to be prepared. The CARVER methodology will assist you in this process.

May you stay safe and vigilant.

General James L. Jones, USMC (Retired)

Introduction

By Luke Bencie

WE ALL REMEMBER WHERE WE WERE ON SEPTEMBER 11, 2001. That horrific day, when those planes hit the Twin Towers and the Pentagon, and the thought of those brave passengers who fought back on United Airlines Flight 93, are burned in our memories forever.

On September 12, 2001, Leo Labaj and I both received phone calls. Leo and I had never met each other, nor did either of us have any idea who the other was. We were both being summoned to work for CIA legend Mick Donahue, recipient of the Trail Blazer Award, given to the top 50 CIA officers who made the greatest impact on the agency during its first 50 years. Mick had been called out of retirement, and his consulting firm, Security Management International (I would later give my firm the same name in honor of Mick), was one of many tasked to help respond to the terrorism crisis facing our country.

On September 14, 2001, when commercial aircraft were allowed to fly again, Leo and I met at Mick's office in Tysons Corner, Virginia. Leo, who had worked with Mick at the CIA and had already retired from a prestigious career of his own as a demolitions expert, was going to be my boss. As a young guy, with minimal experience in counterterrorism, I was grateful to be mentored by men as talented as Mick and Leo. I also had the opportunity to learn from CIA legends Jon Monett and Wil Charrette. But the learning curve was steep, and everyone in America was still on edge. It was truly a baptism by fire for me, career-wise.

Leo and his team (including myself and a few other experts

with military and law-enforcement backgrounds) were assigned to conduct threat and vulnerability assessments of the nation's critical infrastructure. In the days immediately following 9/11, this was of utmost concern. One of our initial projects was assessing a major power distribution center, which if attacked could knock out electricity to several states. During that assessment, Leo taught me the CARVER Target Analysis and Vulnerability Assessment Methodology. Since then, CARVER has become a fixture in my security consulting toolbox.

In the subsequent aftermath of 9/11, Congress passed the Homeland Security Act of 2002, which established the Department of Homeland Security (DHS). While risk and vulnerability assessments were nothing new, it was DHS that issued the 2006 National Infrastructure Protection Plan (NIPP), which outlined the recommended approach for implementing a standardized framework for securing critical infrastructure and key resources (CIKR). Furthermore, the NIPP set-forth the desired roles and responsibilities of sector-specific agencies, such as power, water, transportation, etc., via the Office of Infrastructure Protection (OIP). An over-arching security strategy had finally been established to identify threats and vulnerabilities, while categorizing CIKR utilizing a tiered approach, in order to allow owners and operators to make informed decisions about how best to enhance security.

The new regulations were certainly necessary, but they were not without their challenges. Early DHS vulnerability-assessment standards, as well as those of other U.S. Government agencies, were in some cases too vague, while in other cases too impractical. Implementing the necessary security enhancements to comply with federal guidelines often meant ridiculous amounts of money were being spent unnecessarily on physical security budgets. As is typical of the security industry, there was a knee-jerk reaction of simply throwing money at the problem. CARVER helped resolve the wasteful spending by allowing the assessment teams to slow down and clarify their thinking about threats and vulnerabilities.

Today, Leo and I are still working together. We are also still preaching the value of CARVER as an indispensable tool for counterterrorism and for conducting target analysis and vulnerability assessments of critical infrastructure. Over the past few years, we have conducted dozens of assessments on six continents on a variety of infrastructures – from water and electricity to nuclear plants and stadiums. Now, however, Leo is working for me (but you'd never know it whenever he strolls into my office and gives me his opinions on everything).

In the days since 9/11, CARVER has been propelled to the forefront of the security consulting world. While numerous other vulnerability assessments have emerged and gone by the wayside, the U.S. Department of Homeland Security officially recognized CARVER as the preferred methodology for securing critical infrastructure. The reason: It is time-tested and it works. Also during this period, Leo and I have taught the 3-day CARVER Target Analysis and Vulnerability Assessment Training Course™ to thousands of students from many professions, including the military, law-enforcement and intelligence communities as well as the private sector. We have had numerous Fortune 500 security managers/directors in our courses. In fact, we are one of the only security firms allowed to teach CARVER to foreign governments, per approvals from the U.S. departments of State and Commerce.

It should come as no surprise that Leo has been hailed as the Godfather of CARVER. He is probably the most knowledgeable expert in this field. And though Leo is now an employee of our firm, Security Management International, he will always be a mentor to me. I will never forget the first time I ever performed an overseas assignment. I walked into Leo's office and said, "Okay, Leo, this is my first overseas deployment. Do you have any last-minute advice for me?" His answer, was classic Leo Labaj: "Yeah, two things. One, don't f**k up, and two..." Well, unfortunately, I can't repeat what number two was. But if you know Leo, you understand that his sense of humor is what sustained him through so many years, war zones and harsh living conditions in locations in Africa, Central America and the Middle East.

If you're reading this book, it's probably because you want to gain a greater understanding of how CARVER can help secure your organization's – or your client's – critical infrastructure. If so, you won't be disappointed. There are many assessment methodologies in the security industry, but CARVER is in a class by itself.

As I stated in my book *Global Security Consulting: How to Build a Thriving International Practice*, I have always been amazed at how many risk-assessment and vulnerability-assessment experts are operating out there. I have come to discover, however, that most of them perform such assessments using a basic checklist of security requirements. Items such as "How high is the fence?" "Is there a redundant generator system?" "Is the Continuity of Operations Plan up to date?" And so forth. They might create a good frame of reference, but they aren't useful if you have no practical security experience.

The CARVER methodology is an invaluable tool, but it means nothing without a solid, well-established understanding in the field of security. It would be like giving house blueprints to someone who had never used a hammer, or giving a great recipe to a person who couldn't even boil water – they wouldn't know what to do with it once they got it.

The point is this book was written for those who have a previous knowledge of security assessments and are looking for a way to streamline their findings into both a quantitative and qualitative deliverable format. It will help clarify your thinking and assist with decision-making. It will also provide justification for budgets and resource allocations to internal/external decision-makers and other third parties. In short, CARVER is clear and effective for assessing and ranking assets in order to deliver workable security solutions.

It took several years for Leo and I to put this book together. We hope you enjoy it. May the lessons of CARVER serve you well in your future security endeavors.

Luke Bencie
Tysons Corner, Virginia
May 2018

1. The Origins of CARVER

Why go after tigers when there are plenty of sheep around?
—Hamas training manual

MY MILITARY CAREER (1961-1972) BEGAN IN THE U.S. NAVY as an aviation ordnanceman and in the Connecticut Air National Guard as a bomb disposal technician (my full biography is listed in Appendix IV).

In January 1972, I was accepted into the Central Intelligence Agency at the height of the Vietnam War. Back then, the agency critically needed contract employees with demolition skills to support the Vietnam War effort. I had been assigned to the Special Devices Branch (SDB), under the Technical Services Division, a group heavily involved in the war with a charter for covert action operations and sabotage. That charter dated back to the Office of Strategic Services (OSS), which predated the agency. SDB conducted training in advanced explosive techniques (AET) for members of the special forces and its own paramilitary case officers prior to their deployment to Southeast Asia. The training covered demolition techniques against many types of targets. One training module required of every operative was called target analysis.

During my first six years in SDB, I trained special operations forces and CIA paramilitary units in the use of explosives to conduct or plan sabotage operations. The training included not only how to destroy targets but also how to select targets for destruction to meet mission objectives. The target-analysis methodology of choice at the time was called CARVE.

CARVE had been used long before I arrived at the training facility, so exposure to it was a routine part of my indoctrination, though target analysis was a minor portion of the overall course. In fact, the target-analysis module lasted only two hours, with practical exercises encompassing another four-to-six hours – a total of one day devoted to target analysis and elimination of the selected target.

Although I knew a lot about explosives and could provide valuable training to the attendees, I knew very little about target analysis and CARVE. It became a learning experience for me. I realized from the very beginning, however, that both topics, target analysis and the CARVE methodology, interested me greatly. So, any time I had the opportunity to be the lead target-analysis instructor, I jumped at it.

I learned that CARVE was actually a straightforward process that could be used quickly by operations personnel to select targets. It encompassed five criteria using a value structure of 1 to 5 for each. The criteria were:

Criticality: single points of failure; choke points
Accessibility: ease of access to the target
Recuperability: time needed to restore operations
Vulnerability: operational capability to achieve the desired results
Espy: likelihood that the target will be recognized

The criteria were easily incorporated into a simple table that allowed operations personnel to select targets. Following is an example of a spreadsheet used today to define target selection.

Facility Radar Site	
Date:	
Prepared By:	

ASSET	DESCRIPTION	Criticality	Accessability	Recuperability	Vulnerability	Espy	Target Selection
	Portable Generator	3	4	3	4	4	18
	Control Facility	4	2	3	3	3	15
	Radar Tower	4	4	5	3	5	21

As you can see, the radar tower would be the desired target based on the assessment of the values of the CARVE criteria. It's perhaps oversimplified; nevertheless, it gives an idea of the process.

From 1972 to 1975, as the Vietnam War wound down, CARVE became an important part of the military's target-analysis module. But after the war, and as incidents of international terrorism emerged and increased in frequency, analysts began to find CARVE lacking as a vulnerability-assessment tool.

By 1975, international terrorism was already consuming more and more attention by the intelligence and law-enforcement communities. The phenomenon caused CIA management to consider a reorganization of the SDB in response to incidents such as the following:

- January 24: A bomb exploded in the financial district of Manhattan.
- January 28: Facilities of the U.S. Agency for International Development were bombed in Washington, D.C., and Oakland, California.
- April 19: Four bombs detonated in Manhattan.
- December 23: CIA officer Richard Welch was assassinated outside his home in Athens, Greece, by a group calling themselves Revolutionary Organization 17 November, which eventually murdered 22 individuals in more than 100 attacks.
- December 29: A bomb was detonated at La Guardia Airport in New York City.

It had quickly become evident that bombings and assassinations were going to play a significant role in terrorist activities. In addition to the incidents above, the terror groups became involved in airliner hijackings, hostage-taking and kidnappings.

SDB management determined that its people possessed the skills and capability to address this threat, including expertise in explosives and target analysis. We had a cadre of individuals who knew how to attack a target. It followed, however, that if we knew how to attack a target, we should also be able to prevent an attack on a target. So, we turned our hats around and strove to become counterterrorism "experts." SDB was soon to become SAD, the Special Activities Division. We were beginning to grow.

Counterterrorism

As we explored ways to combat terrorism, we tried to use our experiences from Vietnam. Instead of target analysis, however, we had to develop a vulnerability-assessment methodology that would meet the requirements of protecting the infrastructure of the United States.

We began with CARVE because it was all we knew at the time. We tried the methodology in our vulnerability assessments, but it didn't take long to conclude it wouldn't fit the bill. Some of us got together in a classroom and began defining what comprised a terrorist incident and what the terrorists' goals and objectives were. Our effort became the key to developing the successor to CARVE.

We realized that none of CARVE's criteria quantified the goal of a terrorist attack – we needed something new, and that criteria would be "Effect." Our definition became the impact, severity and consequences of a terrorist incident on the following – but not limited to:

- **Populace**
- **Economy**
- **Media**
- **Politics**
- **Operations**

In 1976 CARVE became CARVER which involved new criteria:

Criticality: single points of failure and degree of importance to the system

Accessibility: ease of access (security system effectiveness)

Recoverability: the time and effort it takes to recover from an adversarial event

Vulnerability: level of exposure to attack based on adversary capability

Effect: scope and magnitude of adverse consequences resulting from malicious actions and the response to them

Recognizability: ability of an adversary to recognize a target and its importance to their operation

We began using CARVER as the operative acronym. We also began testing the methodology conducting vulnerability assessments during many temporary duties (TDYs) around the world. At that time the process was informal and unofficial, so I and other SDB technical officers got together and created what was called the Defense Against Terrorism (DAT) survey.

To employ the DAT survey involved traveling to a foreign country and observing, and studying, its significant infrastructure or assets and determining what vulnerabilities existed based on the threat environment. Even four decades ago, significant worldwide terrorist operations were erupting, and those operations became the catalyst for vulnerability assessments.

To name a few:

- Amman, Jordan: Abu Nidal fighters stormed the Intercontinental Hotel and seized several hostages.
- Kampala, Uganda: Members of the Popular Front for the Liberation of Palestine hijacked Air France Flight 139.
- Karlsruhe, West Germany: Two Red Army faction members assassinated a federal prosecutor and his driver.
- Mogadishu, Somalia: Lufthansa Flight 181 was hijacked by four members of the PFLP.
- Paris, France: Three terrorists fired on EL Al passengers in the departure lounge of Orly Airport.
- Sydney, Australia: A bomb detonated in a garbage bin at the Hilton Hotel. Two garbage collectors and a policeman were killed.
- Washington, D.C.: Twelve gunmen from the Nation of Islam overran three buildings and took nearly 149 hostages.

During the early days of vulnerability assessments, via the DAT survey, CARVER played a crucial role. We needed to quantify the results of our survey to prioritize assets, so appropriate remedial measures could be applied to protecting those assets. Now, vulnerability assessments have become the norm in almost every country of the world, because as we've tragically discovered, no country is immune to terrorist attacks.

I have spent over 40 years of my life protecting people and infrastructure from attack. As such, I am proud to have been a co-creator of the CARVER methodology. Over my extensive counterterrorism career, I have only watched in dismay as the world has become a more dangerous place.

May the lessons in this book assist you in securing your valuable assets.

AS A REPORT ON Critical Infrastructure Protection, issued by the

2. Putting CARVER to Work

Failure to prepare means preparing to fail.

—John Wooden

Government Accountability Office, stated, "our nation's critical infrastructures and key resources (CIKR) – assets and systems physical or virtual, so vital to the United States that their incapacity or destruction would have debilitating impact on national security, national economic security, national public health or safety, or any combination of those matters – continue to be vulnerable to a wide variety of threats. Because the private sector owns the vast majority of the nation's CIKR – banking and financial institutions, commercial facilities, telecommunications networks, and energy production and transmission facilities, among others – it is vital that the public and private sectors work together to protect these assets and systems."

The CARVER methodology we developed in 1976 has been used ever since by the intelligence community, the military and many other government and law-enforcement agencies to help protect our critical infrastructure as well as conduct target analysis against known and potential enemies. It is a time-tested vulnerability-assessment methodology that balances efficiency with reliability. However, before proceeding further, it is important to define what constitutes an effective vulnerability assessment. According to Sandia National Laboratories (from where both authors have received certification in performing vulnerability assessments), a vulnerability assessment (VA) is a systematic evaluation in which quantitative or qualitative techniques are used to predict the physical protection systems (PPS) component performance and overall system effectiveness by identifying exploitable weaknesses in asset protection for a defined threat. After the VA identifies weakness, it is used to establish the requirements for an upgraded PPS design. In addition, a VA is also used to support management decisions regarding protection system upgrades. The "secret sauce" of CARVER, which separates it from other methodologies, is the fact that it offers both a qualitative and quantitative analysis of its findings. This is particularly useful to the security practitioner when he/she needs to explain his/her findings to a non-security decision-maker about implementing solutions and/or discussing budgets. It is important to note that many people might use the terms interchangeably, but we at SMI differentiate between a vulnerability assessment and a risk assessment. For us, a "vulnerability" is a direct weakness that can be exploited at some tactical level, while a "risk" is part of a larger strategic – less well-defined – picture.

As mentioned in the introduction, we derived CARVER from CARVE, a target-analysis methodology that didn't quite work as a vulnerability-assessment tool against terrorist attacks. But when we added the additional criteria, "Effect" and "Recognizability," we turned CARVER into an effective method of gauging the potential impact, severity and consequences of a terrorist incident.

Let's repeat the current six CARVER criteria:

Criticality: single points of failure and degree of importance to the system

Accessibility: ease of access (security system effectiveness)

Recoverability: the time and effort it takes to recover from an adversarial event

Vulnerability: level of exposure to attack based on adversary capability

Effect: scope and magnitude of adverse consequences resulting from malicious actions and the response to them

Recognizability: ability of an adversary to recognize a target and its importance to their operation

Using CARVER successfully requires inserting the proper values into the CARVER matrix. Otherwise, its practitioners won't be able to assess the probability of attack or risk to the asset accurately. When a CARVER matrix is completed, it provides the following results:

- Identifies high risk assets
- Categorizes and prioritizes assets
- Assesses vulnerabilities and consequences

What CARVER does not do is provide risk-reduction recommendations and countermeasures (these will be discussed later in the book and can often be subjective, depending on the level of security expertise of the person conducting/reviewing the vulnerability assessment). The user provides those components based on the values the matrix produces. Here is an example of a CARVER matrix compiled for a water utility with various assets you would expect to find as part of that infrastructure.

Understanding the CARVER Numeric Scoring System

ASSET	DESCRIPTION	DESIGN BASIS THREATS	Criticality	Accessibility	Recoverability	Vulnerability	Effect	Recognizability	RISK	Pa
Water Treatment Plant 1	Chlorine	Hazmat Release	4	3	4	3	5	5	24	0.80
Water Treatment Plant 2	Chlorine	Hazmat Release	4	3	4	3	5	5	24	0.80
Water Treatment Plant 3	Sodium Hydrochlorate	Hazmat Release	4	3	4	3	4	4	22	0.73
Pump 1	Distribution	Physical Destruction	2	3	2	4	2	3	16	0.53
Pump 2	Distribution	Physical Destruction	2	3	2	4	2	3	16	0.53
Storage 1	Fresh Water Reservoir	Contamination	3	3	3	3	3	4	19	0.63
Storage 2	Fresh Water Reservoir	Contamination	3	3	3	3	3	4	19	0.63
Intake	Source Water	Physical Destruction	2	3	2	4	2	4	17	0.57
Well 1	Source Water	Physical Destruction	2	3	2	4	2	3	16	0.53
Headquarters	SCADA/Offices	Cyber / Theft	3	3	3	2	2	4	17	0.57

One of the key elements of the CARVER methodology is the use of an asset numeric scoring system. Assets are identified and evaluated based on the six CARVER criteria, with a score ranging anywhere between 6 and 30 (30 out of 30 usually represents a very vulnerable/high probability of attack asset). Risk to that asset is determined by dividing the calculated CARVER criteria score by the maximum possible score of 30. For example, if the six CARVER criteria in the matrix added up to 15, our risk would be 15 out of 30. To calculate our Probability of Attack (Pa), we would divide the risk number (in this case 15) by our possible score of 30 to get a Probability of Attack score of 50 percent. Note, in our context here of "Risk," we are referring to the risk of losing that one particular asset we are evaluating – not the risk to the entire system. It must be remembered that the higher the risk number, the more favorable the chances for the adversary.

The below chart defines the numeric values for the six CARVER criteria.

Criticality

Criticality is based on the importance of the asset. An asset is critical when its destruction or compromise would significantly impact the output, mission or operation of the asset, facility or infrastructure. Its elements include:

- Single point of failure
- Choke point

Accessibility

Criticality Value Structure	Scale
Loss would be mission-stopper	5
Loss would reduce mission performance considerably	4
Loss would reduce mission performance	3
Loss might reduce mission performance	2
Loss would not affect mission performance	1

Accessibility is the ease with which an asset can be reached directly or indirectly by an adversary to accomplish a predefined mission. Its elements include:

- Transportation
- Security
- Physical barriers
- Security manpower
- Law enforcement
- Levels of detection
- Timeliness of response

Recoverability

Accessibilty Value Structure	Scale
Easily accessible. No effective security	5
Accessible	4
Somewhat accessible	3
Difficult to gain access	2
Very difficult to gain access	1

A measure of the amount of time required to replace, repair or bypass the destruction or damage inflicted to an asset, facility or infrastructure. Its elements include:

- Sources of components
- Age of components
- Spare parts
- Ability to bypass operation
- Ability to cannibalize

Recoverability Value Structure	Scale
Extremely difficult to replace. Long down time	5
Diffcult to replace with long down time	4
Can be replaced in a relatively short time	3
Easily replaced in a short time	2
Immediate replacement. Spare parts are readily available or asset redundancy	1

Vulnerability

The ability of the adversary to damage, destroy, compromise or exploit the asset using available capabilities. Its elements include:

- Nature and construction of the target
- Degree of damage required
- Available assets or materials
- Training
- Adversary capability

Vulnerabilty Value Structure	Scale
A dedicated adversary has the capability and expertise to attack	5
A dedicated adversary most likely has the capability and expertise to attack	4
A dedicated adversary may have the capability and expertise to attack	3
A dedicated adversary most likely does not have the capability and expertise	2
A dedicated adversary does not have the capability and expertise to attack	1

Effect

The scope and magnitude of adverse consequences that would result from malicious actions and responses to them. This context addresses all significant effects, whether desired or not, which might result if the selected asset is attacked

Vulnerabilty Value Structure	Scale
Very high sociological, economic, political impact; considerable loss of life or injuries	5
High impact; some loss of life or injuries	4
Moderate impact; some adverse impact on persons	3
Little impact; no adverse impact on persons	2
No unfavorable impact	1

Recognizability

The degree to which a system or asset can be recognized as a desirable element meeting the goals of the adversary's objectives. Its elements include:

- Available information (internet, open source, etc.)
- Public exposure
- Physical appearance
- Terrain
- Active signatures

Vulnerabilty Value Structure	Scale
Easily recognized by all with no confusion	5
Easily recognized by most	4
Recognized with some training	3
Difficult to recognize. Confusion probable	2
Extremely difficult to recognize without assistance	1

In the final chapter, you will see all of these value structures combined into a single chart. In the following chapters, we will be more specific about how CARVER is used in a variety of vulnerability assessments. We will discuss critical infrastructure such as water and power utilities but will not be limited to those. There are many vulnerability assessments, and each will be discussed on its own merits.

CARVER and EPA'S Mandated Water Vulnerability Assessments

In CARVER's early days, we would define the risk to assets by using whole numbers. As in the CARVER matrix above, the risk is defined for each asset such as Water Treatment Plant 1 with a Risk Value of 24. It wasn't until 2003 that the Probability of Attack (Pa) was also quantified and inserted into the CARVER results. This was done primarily to solve issues in another vulnerability-assessment methodology that was mandated by the Bioterrorism Preparedness and Response Act (BPRA) of 2002, which required that drinking water utilities serving more than 3,300 people conduct vulnerability assessments. The Environmental Protection Agency (EPA) provided grant money to those utilities serving populations of more than 10,000, but the agency allotted no grant money to the smaller utilities. For those systems serving between 3,300 and 10,000 people, EPA created and published the Security Vulnerability Self-Assessment Guide. The effort proved inadequate because the assessments were completed by non-security personnel.

The assessments were meant to determine the security of U.S. water systems and dams as well as Supervisory Control and Data Acquisition (SCADA) systems, which provide command-and-control over day-to-day operations. In turn came development of the threat assessment and design basis threats (DBTs) for the water industry. Once those tools were available, we could perform target analyses and/or vulnerability assessments.

Then, a discovery of water-infrastructure information in terrorist hands strongly suggested that the U.S. system could be vulnerable to attack. As a result, following the creation of the BPRA, the EPA contracted with Sandia National Laboratories to develop the Risk Assessment Methodology for Water, commonly known as RAM-W. Sandia also created an assessment methodology for dams known as RAM-D. EPA then required RAM-W and RAM-D for all applicable assessments, as well as training in their use.

Prior to doing my first water assessment, under the new mandate, I attended a training class in San Diego on RAM-W. I was the only person in the room with a security background. The rest of the attendees were engineers and/or water-department employees. That meant a lot of them lacked previous experience conducting security vulnerability assessments, and therefore many of their employers had to hire security experts to partner with them.

Following training, I conducted a water vulnerability assessment in Southern California using RAM-W. Afterward, as required for every such assessment in the United States, I wrote a report for the utility and EPA, which used this report to follow up with the water authority to ensure they adopted the report's recommendations and countermeasures.

You can see a diagram of the RAM-W process below. I have no problem with RAM-W as a whole; only with the formula that quantifies risk, one of whose key attributes is Pa or Probability of Attack. RAM-W creators apparently thought they could not quantify the probability of attack, so they used a constant of "1" in the formula. The problem was this meant all assets had the same Pa – in other words they were all considered perfect targets, which is absurd. Based on security system effectiveness and other criteria, that number must decrease and be represented by a decimal less than 1.

Another problem with this formula, as written, was it did not prioritize assets enough, so remedial action to solve security issues

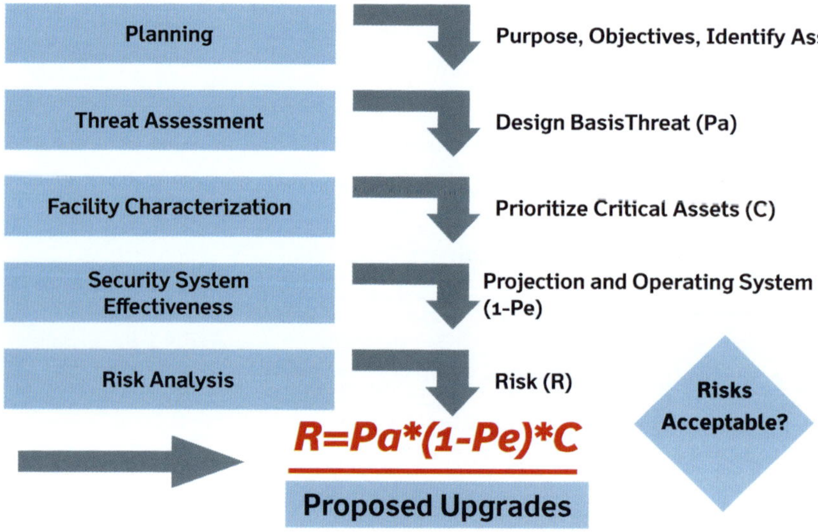

could not be taken based on the utility's resources. Some utilities would have 50+ assets and, if Probability of Attack was not clearly defined, all assets would be assigned the same risk. The problem was that EPA would then require a utility to spend resources to reduce the risk on all assets instead of those most attractive to the adversary.

After the first vulnerability assessment using RAM-W, we quickly detected the problem: We could not differentiate among assets. Nevertheless, we were stuck with RAM-W because EPA had mandated its use. But that didn't mean we couldn't fix the formula so it provided realistic information. Thus, we incorporated CARVER to define the Probability of Attack. As you can see in the matrix below, it easily defines Pa. You can also see, by applying the CARVER methodology to RAM-W, you can more adequately determine the risk to various assets.

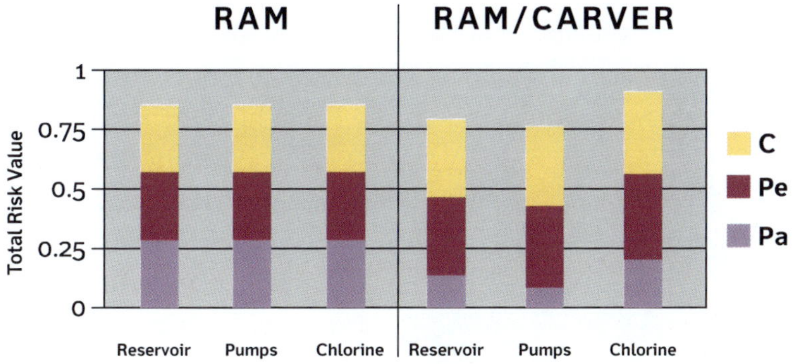

In RAM-W the Probability of Attack is the same for every asset because of the constant of 1. By inserting CARVER values, you can easily prioritize assets based on risk – allowing the organization involved to apply its limited resources to reducing the risk of the most attractive assets to an adversary.

Case Study 1: A Water Utility Assessment

Less than 60 days after the horrifying events of 9/11, I was called upon to conduct a full vulnerability assessment of a large municipal water utility in Washington state. This was the first time I had the opportunity to conduct such an assessment, so the first one was going to be a learning curve for me.

Luckily, the engineering and operations staff at the utility took me under their wing and taught me everything they could about their system and how adverse conditions could affect it. At the time, EPA had not mandated an assessment methodology, and we did not know when one would be coming down the pike. Therefore, my first vulnerability assessment involved using the CARVER methodology, which provided qualitative and quantitative results that allowed the utility to identify vulnerabilities, prioritize its assets and initiate countermeasures.

At the time, I had used CARVER in a multitude of vulnerability assessments worldwide. This opportunity with the water industry did not cause me concern other than learning the system and applying CARVER to the information I gathered.

It would be another two years before EPA got involved in the vulnerability-assessment process. But I want to remind the reader that the CARVER methodology was enormously successful as a tool to prioritize assets, while prioritization was not effective using the EPA-mandated process. The threat spectrum I have provided is typical for most water utilities, and as you can see a hazmat release (chlorine, for example) has the most impact but is least likely to happen. This particular utility purchased finished water from a nearby utility, thereby negating the requirement for a water treatment plant using chlorine. Once you remove hazmat release as the highest impact event, the one that follows is contamination of the fresh water supply.

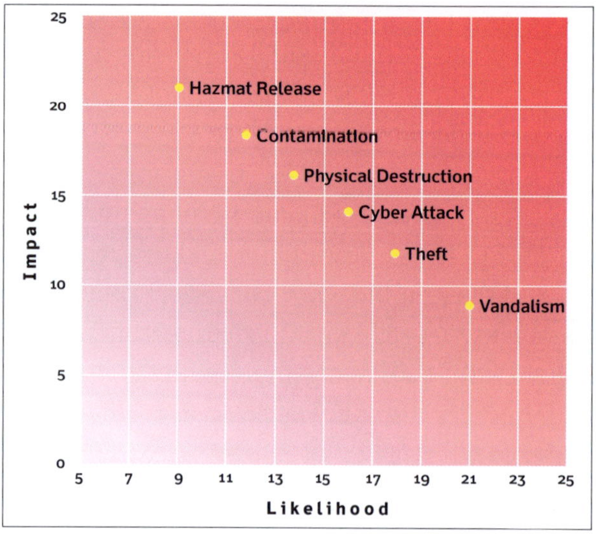

Contamination, like hazmat release, will shut down system operations for some time. One of the concerns of any water utility is the possibility of contamination of the fresh-water supply. It was determined that fresh-water storage and distribution systems were the only possibilities for contamination. Contaminating the source water prior to disinfection is not a viable attack methodology. So, it followed that much of our recommendations to deny a contamination event would be applied to fresh-water storage and distribution systems. Fresh-water storage reservoirs were the most vulnerable to contamination and attack; however, these vulnerabilities were easy to fix, and that fix would prevent an adversary from gaining access to water via the use of locks and various other delay mechanisms. If by chance an adversary did gain access, an alarm would go off, activating an automatic system shutdown – which would take that reservoir out of the system and prevent contaminated water from entering the distribution system.

Protecting the distribution system is inherently much more difficult than protecting a fresh-water reservoir. One of the CARVER criteria was accessibility, and this criterium played a role in defining asset prioritization. There are so many access points in the system where a contaminant could be introduced, and in many cases they are totally unprotected. Nevertheless, contaminating the distribution system is difficult because the adversary must introduce a contaminant by overcoming the pressure within the system. This would require the adversary to employ specialized equipment such as pumps to introduce the contaminant. That said, if the target has enough value, the difficulty of access to the system might be warranted. Examples could be VIP residences, hospitals, food processing and/or soft-drink manufacturing and packaging facilities. All became important in the vulnerability assessment.

In one case, there was a vault located alongside a public road that contained pipes providing water to a very influential person who could be a potential target. This well-known celebrity had a security force both on the property, and at the vehicle-access point. But the vault, located just outside the property, was completely unobservable by the security force. A small team using counterfeit or stolen uniforms from the water utility and, possibly, magnetic water department signs on the side of their van, would have a cover for action to inject contaminant into the distribution system leading to the residence. A simple countermeasure, which was recommended and which would prevent this attack, was CCTV surveillance of the vault. This was installed and worked extremely well, allowing any maintenance or work done on the vault to be confirmed while in progress with the water utility.

Case Study 2: A Power Distribution System

I have had the opportunity to conduct dozens of critical infrastructure vulnerability assessments over the past two decades. These have ranged from water utilities, sports stadiums, military bases, power grids, universities, office buildings and transportation systems. On a few occasions, we assessed a power-distribution system, which is essentially a central clearing hub where electrical power is bought/sold and then distributed across a geographic region. If an adversary were to "take out" one of these centers, it could mean a temporary power blackout across several states (a terrorist's or hostile foreign government's dream). Therefore, to ensure continuity of daily public- and private-sector operations (not to mention everyday life), a hardened physical profile at this type of facility – with the ability to detect, deter, delay, respond and mitigate – is paramount.

Utilizing the principles of the CARVER methodology, I, along with other former CIA officers and FBI explosive experts, determined how best to disrupt the flow of power at one of these facilities with a simple yet direct clandestine attack. The methods undertaken were part of standard tradecraft principles, which were learned through real-world operations and are still in practice today within the intelligence community. I cannot disclose the technique used to exploit this particular vulnerability, but it was nonetheless identified and was quickly rectified to ensure further protection.

The point is that a standard security vulnerability assessment checklist would have never identified this type of exploitable opportunity for sabotage. Only through a comprehensive understanding of how the CARVER methodology works for operational planning, does the assessor begin to reframe his/her thinking to identify and take advantage of hidden vulnerabilities within an infrastructure. Furthermore, it demonstrates the need for organizations to incorporate third-party subject-matter experts to look at their critical infrastructure through an impartial lens, rather than use in-house employees who might be blinded to vulnerabilities simply from day-to-day exposure.

The CIP-014 Department of Energy Mandate

Most of the U.S. power grid is owned and operated by the private sector, so the Department of Energy chartered the North American Electric Reliability Corporation (NERC) to monitor the reliability and security of the nation's power grid. NERC created a new reliability standard for Critical Infrastructure Protection (CIP) to ensure the grid remains robust and resilient.

NERC promulgated the regulation known as Critical Infrastructure Protection 014 (CIP-014) to ensure that individual power companies comply with a standard level of security protection. The CIP-014 regulation states:

> *To identify and protect Transmission stations and Transmission substations, and their associated primary control centers, that if rendered inoperable or damaged as a result of a physical attack could result in a widespread instability, uncontrolled separation, or Cascading within an interconnection.*

As Director of Critical Infrastructure Protection for Security Management International (SMI), I frequently employ CARVER to evaluate assets within the grid. In fact, since the CIP-014 requirement was enacted, I have trained many in the energy sector on how to use CARVER to conduct their own assessments.

NERC CIP-014 Requirements

The CIP-014 standard is non-prescriptive; that is, it does not mandate specific security measures. But the intent of the standard is clear in expecting electric utilities to step up their game when it comes to protecting the power grid. Utilities have a strong incentive to comply with this standard, because NERC has the authority to levy fines for noncompliance of up to $1 million per day.

CIP-014 consists of six primary requirements that transmission owners and operators must follow to enhance the physical security of identified transmission substations and primary control centers. It starts with conducting an initial risk assessment to identify critical assets, performing a vulnerability analysis of potential threats, and developing associated security plans to mitigate the risks identified. Last, the utility must enlist an independent third party to review and validate the plans. The six requirements outlined in CIP-014 include:

R1: Perform an initial risk assessment to identify critical facilities (transmission substations and primary control centers)

R2: Have an unaffiliated third party review initial risk assessment conducted under R1

R3: Notify any transmission owner that is not under operational control of a recognized station that the asset has been identified during R1

R4: Conduct an evaluation of the potential threats and vulnerabilities of a physical attack to identified assets under R1

R5: Develop and implement a documented physical security plan that addresses the site-specific threats and vulnerabilities indicated in R4

R6: Have an unaffiliated third party review the evaluation performed under R4 and the resulting security plan developed under R5

CIP-014 gives utilities the choice to conduct these actions utilizing in-house security professionals, or outsource the process to an independent third party. Regardless of who performs the work, however, the process used to conduct the threat and vulnerability assessment and physical security plan (Requirements R4 & R5) must be well-documented and defined, so it can withstand potential scrutiny from a variety of sources such as concerned customers, local governments and public utility commissions, as well as the media.

Employing CARVER to Meet the CIP-014 Mandate

To best meet CIP-014 requirement R4, we employ CARVER. We find it an extremely well-suited tool that can be used both to evaluate the potential vulnerabilities of an electric grid and prioritize critical assets based on their probability of being attacked (i.e., target attractiveness). The CARVER assessment encompasses the following basic principles:

- Examine all security programs from the terrorist's or adversary's viewpoint (i.e., how would I, if I were the adversary, attack this asset and what would I target?).
- Determine what countermeasures would be required to prevent a terrorist or adversary from achieving success in launching and completing the operation.
- Assume that conspicuous security procedures and equipment will not stop a determined terrorist or other adversary.

Because CARVER addresses both the physical security of a facility and the operational capabilities of terrorists or other adversaries, the methodology acts as a target-selection tool that can identify and prioritize the targets most attractive to attack by an adversary. CARVER helps to determine the relative risk for each threat and target combination using algorithms to analyze potential consequences, relative importance of targets to the aggressor and security vulnerability.

Based on a CARVER assessment, we can recommend countermeasures to minimize the risk and, subsequently, reapply the methodology to determine if risk reduction has been achieved. We then continue the process until we can identify the most cost-effective method of reducing the risk to an acceptable level.

Since we began using the new requirements of NERC CIP-014, we have not only helped electric utilities comply with Department of Energy standards but also have documented both the quantitative and qualitative processes used to review all of the critical assets onsite. Furthermore, CARVER provides a clear, quantifiable (and repeatable) process that can effectively document and defend the decisions made to identify and protect critical assets.

I have described R4 and R5, which is the security-assessment and security-implementation plan, but I also want to mention R6. This is an independent review by a third-party regardless of whether that third-party conducted R4 and R5 assessments. I and SMI have conducted successful R6 reviews of a number of utilities. In many cases this involved rewriting and/or editing R4 and R5. This was done primarily for consistency throughout the reports.

CARVER: The Counterintelligence Metric

As I hope you can see, CARVER is a versatile methodology. It can be applied either as a target-analysis or vulnerability-assessment tool to protect assets other than physical. Further examples of this will be presented later in the book. During my career, I used CARVER to perform counterintelligence vulnerability assessments for military installations concerned about espionage, and in doing so I helped identify remedial measures to protect those assets. All it required was changing the definitions of the criteria to meet the requirements of the counterintelligence assessment. As such, the assessment was based on information assets.

Following are the definitions for each of the CARVER criteria, which include those for both security and counterintelligence assessments.

Some explanation might be useful for those who might not have an intelligence background. For example, Critical Program Information, or CPI, usually refers to military programs. Special Access Programs, or SAPs, limit access to those with the Need to Know. And FIS stands for Foreign Intelligence Services, which in the case of counterintelligence assessments are your adversaries.

Unlike a physical assessment performed on infrastructure, a counterintelligence vulnerability assessment focuses on, but is not limited to:

- OPSEC policy and procedures
- Human resources & security clearances
- Information technology, internet, email, websites
- Cameras, recorders and laptops restrictions

(CONTINUED ON PAGE 38)

CARVER	Security Assessments	Counterintelligence Assessments
Criticality	Identify critical systems, single points of failure or choke points	Identify CPI, SAP, Sensitive Classified or Unclassified information critical to national security, the economy or technological lead
Accessibility	Determine ease of access to critical systems, assets, operations	Determine ease of access to information by Foreign Intelligence Service (FIS) through typical methods of espionage
Recoverability	The time it would take to replace or restore a critical system against maximum acceptable period of disruption	The time it would take to recover from the loss of information
Vulnerability	Assess assets' risk to selected adversary capabilities and methods of operation	Vulnerability of the information based on FIS strategic capabilities and sources and methods
Effect	Consider scope and magnitude of adverse consequences that would result from malicious actions and responses to them	Consider scope and magnitude of the loss of information to national security, economics, or technological lead
Recognizability	Evaluate likelihood that potential adversaries would recognize that a system was critical	Evaluate likelihood that potential FIS would recognize that the information exists, where it physically resides and who has access

- Travel policies
- Workforce/ethnic breakdown/foreign vs. u.s., etc.
- Open source exposure of unclassified information and key employees
- Presentations, papers
- Physical security
- Visitors, foreign or U.S citizens
- Technology transfer and cooperative research & development agreement

SMI prides itself on possessing the skills base and knowledge to conduct counterintelligence vulnerability assessments using the modified CARVER methodology. In fact, SMI is uniquely situated to conduct any vulnerability assessment against any assets through modification of CARVER to fit the individual case requirements. This is important, because the experts at SMI did not just learn CARVER from a book. They have used the the methodology during lifelong careers in security and intelligence. Chapter Six is dedicated to how CARVER can be utilized as part of an intelligence targeting campaign.

Summary of the CARVER Experience

Since 1976, I have conducted hundreds of vulnerability assessments using CARVER in 50-plus countries around the world and, more recently, since the 9/11 terror attacks I have conducted vulnerability assessments involving much of the critical infrastructure in the United States. In many cases, CARVER allowed me to improve the results of previous assessments. Some examples:

Water infrastructure. A water utility in Southern California asked me to review a vulnerability assessment conducted by another entity that had inadequately prioritized their assets. Using CARVER to supplement the methodology they had employed, I identified several gaps in the results. CARVER reduced the number of assets requiring remedial action from 76 to 7.

Power grids. Like water utilities, power grids have diverse assets that require protection. I applied CARVER to several previous assessments that had passed the facilities with flying colors. I found vulnerabilities the assessments had missed.

Sports stadiums. These facilities are unique in that they are open to the public – including adversaries. This makes vulnerability assessments difficult because of the significant challenge in devising appropriate countermeasures. CARVER has excelled in this regard because it addresses unique threats to sporting events and other large public venues.

Military bases. I have conducted physical and counterintelligence vulnerability assessments on military installations both domestic and overseas, and I found many foreign installations had no methodology to speak of to conduct physical vulnerability assessments. I introduced them to CARVER, and to this day they are still using it. CARVER has a distinct advantage: It is very, very easy to use.

Chemical facilities. Chemical installations are also unique because they all fall under tremendous oversight by government agencies. Chemical companies must protect their installations so no harm comes to anyone who might live in surrounding areas. Not only are the standard safety issues enormous but there is also the potential of an attack by a variety of adversaries. When I introduced CARVER to many of these companies it was well received. My standard approach was to give a two-day course immediately followed by an onsite vulnerability assessment, which I monitored, and writing the report, which I supported.

Airports/seaports. Most of my experience in this area has been international, particularly during the late 1970s and into the '80s. During this time, international terrorism was emerging overseas and even in the United States. Domestically, however, the country had a security infrastructure in place to conduct vulnerability assessments. Not so, internationally. I used CARVER extensively for these infrastructures, and it functioned very well. Again, CARVER is easy to teach, easy to use and effective.

Office buildings. These facilities present unique challenges for conducting vulnerability assessments; even more so when it comes to applying countermeasures. That's because threats to office buildings can vary greatly from terrorism to vandalism. Moreover, visible security measures are generally unwanted. Nevertheless, I have found that CARVER provides a useful matrix to establish the risks to all assets within a building, including the staff.

Shopping malls. Like other infrastructures that are open to the public, the threat of terrorism and other adversarial activities are real and constant. Yet mall management companies tend to opt out of doing vulnerability assessments, which could entail installing visible countermeasures. In many cases management doesn't want to know where they are vulnerable, because such knowledge requires doing something about it. Still, I have been able to apply CARVER to some malls to demonstrate their vulnerabilities.

3. Threat Assessment and Design Basis Threats

When we have a free path, we go forward. If we meet an obstacle, we go around it. If the object cannot be overcome, we retreat. When the enemy is unprepared, we surprise him. If he is alert, we leave him alone

—Bader-Meinhoff Gang
(German urban terrorist organization)

Threat Assessment – A Vital Step in the Vulnerability Assessment Process

Historically speaking, those responsible for protecting our nation's critical infrastructure and other vital facilities have presumed that a major attack would most likely occur in cities near our borders with Canada or Mexico; cities such as Seattle, San Diego and El Paso, which offer greater ease of access. Or, would-be terrorists might target our largest metropolitan areas such as New York City, Los Angeles, Chicago, Washington, D.C., San Francisco and Boston, where the notoriety and impact of an attack would be (and have been) greatly amplified.

But the destruction of the Murrah Federal Building in Oklahoma City on April 19, 1995; the mass-shootings at Fort Hood, Texas, and the Pulse nightclub in Orlando, Florida, and even the Boston Marathon bombing have shown that holding such presumptions can be dangerous.

It has become obvious to many observers that terrorists will attack targets of opportunity. Typically, they look for easy prey: locations and gatherings that are less secured and manifest exploitable weaknesses (e.g., Orlando and Las Vegas mass shootings). As our larger cities and urban areas, often with federal and other financial assistance, take meaningful steps to eliminate their vulnerabilities and improve security, their attractiveness as targets declines considerably. Regardless of size or location, the threat of an attack is greatest where the defenses are weak and attention to security is lax.

It is therefore critical that the information we gather in a threat assessment can be used in a vulnerability assessment as well to evaluate an asset's vulnerability to potential threats, the effectiveness of existing security measures against such threats, and the changes necessary to minimize the threats and/or the consequences thereof. Threat-assessment information also can be used to develop design basis threats (DBTs) and a threat spectrum appropriate to an asset or assets.

In order to develop the threat spectrum for a critical infrastructure, we must characterize the potential adversaries, which will vary according to targeting the facility itself or targeting a remote asset of the facility. Their method of attack will also vary greatly based on an adversary's capabilities. Some examples:

- International and domestic terrorist organizations
- Hate groups, militias, cults, radical environmentalists and opposition movements
- Vandals and computer hackers
- Disgruntled employees, contractors and other insiders
- Disgruntled customers
- Criminals

We divide the threat assessment into sections according to the goals and objectives of the vulnerability assessment. Simply stated, the threat assessment is the appraisal of collected intelligence and other available information regarding the spectrum of threats to operations, programs, facilities and people.

Potential Adversaries

- What groups or persons are most likely to pose a threat to the target?
- Why are these groups or persons a threat?
- What types of attacks have occurred in the past and might happen again?
- What are the objectives or motivations of the potential aggressors?

Capabilities

- What are the capabilities of the potential adversaries?
- What is their access to hazardous items?

Mode of Attack

- How are the threats likely to be carried out?
- What is the likelihood that the threats will be carried out?

Potential Consequences

- What are the likely effects of an attack?
- What is the expected response time?

International Terrorism Threats

The catastrophic events of 9/11 marked a milestone in international terrorism, but they by no means constituted an isolated group of incidents, as horrific as that day turned out to be. Subsequent years saw the Madrid train bombings, the London train and bus bombings, the attacks in Mumbai and, more recently, the attacks in Paris, Brussels and Manchester. The planning, execution and magnitude of most of those attacks exceeded pre-9/11. Now, global terrorism has become an unbounded threat, utilizing more aggressive, horrific and deadly tactics. Domestic terrorism has equally become an unbounded threat as characterized by the Fort Hood, Aurora, Sandy Hook, Orlando, San Bernardino, Las Vegas and other active-shooter incidents, as well as the Times Square attempted bombing and the Boston Marathon bombing.

Terrorist Organizations

Many terrorist groups are well-organized, well-trained and well-funded, and their members are experienced, dedicated and prepared. They conduct extensive research on their targets, identify their weaknesses, and plan their attacks. We know that members of al Qaeda, other terrorist groups and so-called lone-wolf extremists often receive training in the use or conduct of:

- Firearms, explosives and incendiary devices
- Biological agents, chemical substances, hazardous materials and other contaminants
- Hand tools, small power tools and other similar equipment

- False identities, disguises and clandestine operations
- Reconnaissance, casing, surveillance, photography and observation

Threat Spectrum – Mode of Attack

An attack against our critical infrastructure/key resources (CI/KR) could take one of the following forms:

- Physical destruction of an asset not associated with an explosive device
- Theft of property from the facility, parking lots, loading docks, etc.
- Improvised explosive devices (IEDs) or vehicle-borne improvised explosive devices (VBIEDs)
- Ballistic attack with the goal of multiple casualties or damaging equipment
- Vandalism of a facility or asset, including defacement and damage of property
- Cyber attack

If we take each of these potential modes of attack, and plot their impact and likelihood on a graph, we can see that the most likely events are theft and vandalism, although they cause the least impact on facility operations. That said, it is imperative that critical infrastructure and key resources protect themselves, particularly against the high-impact types of attacks, which usually cause death and injuries.

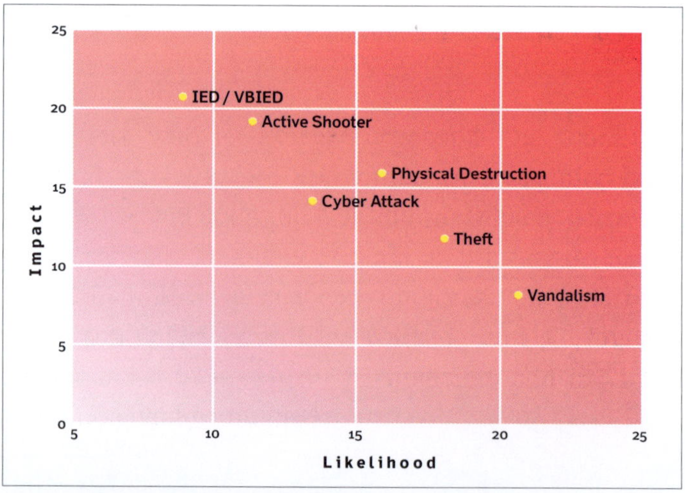

Available information and recent experiences suggest that critical infrastructure or key resources could become the target of a variety of adversaries. Terrorists and others seek out unprotected or "soft" targets; unfortunately, most of America's critical infrastructure currently qualifies as such. Authorities that fail to take proactive measures to evaluate vulnerabilities, shore up their security systems, implement threat detection measures, update their emergency operations plans and train their employees to handle incidents to reduce the likelihood of an attack and/or reduce the severity of its consequences leave their facilities open to attack.

Design Basis Threats (DBTs)

If you need to design and evaluate security systems and programs that will effectively protect critical infrastructure, then you must first understand the methods and capabilities of potential adversaries. Using the information developed in the threat assessment, and applying our experience in conducting vulnerability assessments for a wide range of assets, we have provided examples of design basis threats (DBTs) as some of the potential attributes which might affect critical infrastructure and key resources. The DBT assessment identifies adversaries, goals and capabilities that might include weapons, explosives, tools and tactics that are useful in determining the deterrence, detection/assessment, delay and response criteria for the design and evaluation of an effective physical security protection system.

Deter: A countermeasure strategy intended to discourage a breach of security by means of fear or doubt. The countermeasure can be real or imaginary.

Detect: A countermeasure strategy intended to provide identification of inappropriate, incorrect, malicious or anomalous activity by an adversary attempting to breach security.

Delay: A countermeasure strategy intended to provide various barriers to slow the progress of an adversary until an adequate response can be devised that prevents an unwanted occurrence.

Assess: Part of the detection process used to evaluate, judge or measure any possible adversarial activity.

Communicate: Part of the detection process that transmits alarms and video images to a central location where they can be assessed and recorded.

Respond: A security strategy that reacts to an adversarial activity after detection and assessment. It can include notification of authorities, evasive action, confrontation and apprehension.

In the absence of a sufficiently detailed description of the threat, it is difficult to determine precisely the amount of protection required to protect a given asset from the unacceptable consequences of an attacker. Thus, we create design basis threat assessments that define the characteristics and capabilities of an external or internal attack or an attempt to otherwise compromise an asset. The design basis threat then becomes the roadmap for the vulnerability assessment. Generally, we develop the DBT and provide it in paragraph form, including a description of the following attributes of the perceived threat.

- Adversary description and capability
- Type equipment an adversary would use
- Adversary knowledge
- Tactics/MO
- Adversary goals

Examples of Design Basis Threats

1. Domestic/International Terrorist(s) – Large Vehicle Borne Improvised Explosive Device. An attack planned by 1-2 outsiders with access to, knowledge of and/or experience with explosives. They will use sophisticated hand tools, power tools, firearms and have the other resources necessary to obtain and/or make a large vehicle borne improvised explosive device (LVBIED). Their goal

is to utilize an LVBIED against a particular target to kill and injure numerous victims and cause significant destruction of a facility's infrastructure, denying its use for a lengthy period, thereby causing serious economic consequences for the community. The attackers will harm anyone hindering their efforts.

2. Domestic/International Terrorists – HAZMAT Release. An attack planned by 1-2 outsiders with no authorized access but knowledge of chemical manufacturing systems obtained through open-source information, observation and solicitation. They will use sophisticated hand tools, small power tools and firearms. They will be trained in the use of explosives and/or incendiaries. Their goal is to destroy chlorine tanks to vent chlorine gas to the atmo-

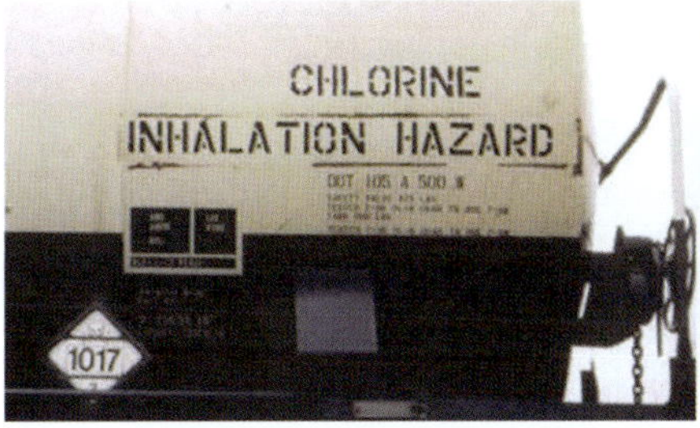

sphere. Their purpose is to cause panic and death within a local geographic area. They will harm employees if necessary.

3. Outsider – Vandal. An attack planned and conceived by a single outsider without authorized access or inside information using unsophisticated hand tools or small power tools with the intent of stealing, defacing or damaging assets of the facility. The outsider does not intend to cause physical harm to facility employees or end-users.

4. Outsider – Cyber. An attack planned and conceived by a single hacker without authorized access or inside information attempting to use a SCADA or other automated system to remotely control various assets and/or deny access to such systems or assets. The outsider does not intend to cause physical harm to utility employees or end-users.

Creating a threat assessment and DBT is Step 2 of the vulnerability-assessment process. Step 1 of the process is creating an inventory and identifying the assets to be protected. That is essential to creating the DBT for each individual asset. Once created you're off and running for your site visits. DBT can be summarized in the following table.

THREAT ANALYSIS SUMMARY

Facility: Anywhere Chemical Processing Corp.

Adversary	Number	Equipment/ vehicles	Knowledge	Weapons	Tactics
Outsider-terrorist	1-2	Hand tools/car, van, explosives	Special operations training	Handguns	Explosive attack against chlorine tanks
Outsider-terrorist	1-2	Hand tools/pumps car, van	special operations training	Handguns	Contamination of processing operations
Insider-disgruntled employee	1	None	SCADA operator	None	Corrupt SCADA ops
Outsider-hacker	1	None	Hacking skills	None	Denial of service

The information for the summary is extracted from the DBT that were written in paragraph form.

In the subsequent chapters, I will define the target-analysis and vulnerability-assessment process.

Threat Assessment - Terrorist Bombs and Bombings

I want to separate this section from the threat-assessment chapter above, which is general in nature. The following is more specific because the threats from terrorist bombings can be horrific as in the case of Oklahoma City in April 1995. That bombing, which was considered domestic terrorism, cost 168 lives; however, I will focus on international terrorism in this section.

Bombings and attempted bombings are on the rise worldwide, but attacks in the United States have been limited. Why? We have created a more stable security environment.

- Police/security forces training, techniques and equipment have improved.
- New countermeasures in place since 9/11 have created the perception of difficulty in operating in the United States.
- Advanced, counter-IED technology has emerged; nevertheless, bombings have been attempted against U.S. targets since 9/11.

Attempted bombings against targets in the United States since 9/11 have included:

- December 2001: Richard Reid, the infamous shoe bomber, attempted to destroy American Airlines Flight 63 from Paris to Miami wearing plastique explosives packed into the soles of his shoes.
- June 2004: The Columbus Shopping Mall bombing plot was a plan to blow up an unnamed mall in Columbus, Ohio. The plot was disclosed by federal authorities on June 14, 2004, when an indictment was brought against Nuradin Abdi. Abdi was part of a clandestine cell of al-Qaeda.
- September 2009: Hosam Maher Husein Smadi, a citizen of Jordan, was arrested on suspicion of planning a terrorist bombing of Fountain Place, a downtown Dallas, Texas, skyscraper.
- December 2009: The so-called underwear bomber, Umar Farouk Abdulmutallab, attempted a suicide bombing of Northwest Airlines Flight 253.
- May 2010: Two Pakistani men attempted an evening car bombing in crowded Times Square. The device was activated but failed to detonate.
- October 2012: Ahsan Nafis, a Bangladeshi, was sentenced to 30 years in prison over a failed attempt to detonate a truck bomb outside the U.S. Federal Reserve Bank in New York City.

All of the above incidents failed, but sooner or later the bad guys win, as in the case of the Boston Marathon bombing on April 15, 2013, where two bombs packed inside pressure cookers killed three and injured more than 260. The ensuing manhunt resulted in the death of an MIT police officer and suspect Tamerlan Tsarnaev, and ending with the capture of Dzhokhar Tsarnaev.

The bomb traditionally has been the favorite weapon of the terrorist; however, vehicles are finding more and more use as terrorist weapons. There is a reason for this: Military-grade explosives and detonators are getting harder to find, steal or make. As was in the case of a recent failed bombing on a subway in London perpetrated by an alleged 18-year-old, an individual who did not construct a workable improvised explosive device. Instead of a high-order detonation, he primarily got an incendiary effect.

Why the Bomb?

Thc target, nature and size of the bomb is limited only by the bombmaker's imagination; it becomes a function of their access to the target and to the bomb-making materials. There is a lot of information on the Internet that can provide instruction on how to build an improvised explosive device. One such example is *The Anarchist Cookbook.* This publication has been around for decades, but some of its instructions will fail or maybe kill or injure the bombmaker. Other examples of bomb-making exist online, and it takes only a simple Internet search to find the information. One of the most notorious is Inspire magazine. Published by jihadists, it describes bomb-making and bombing operations.

Bomb Threat Planning

In bomb-threat planning, the adversary must consider what the target is and then choose from the variety of devices and methods of attack. These can be considered in more detail when characterized in general categories such as:

- Large-scale, vehicle-borne vs. hand-carried IED
- Suicidal vs. non-suicidal attack
- Time delay vs. command detonated

The next step of the operation is to acquire the explosive material. Acquiring military explosives and detonators is relatively difficult; thus the proliferation of homemade improvised explosives such as:

- Ammonium nitrate-based
- Urea nitrate-based
- Chlorate-based
- TATP (tri-acetone tri-peroxide)
- HMTD (hexamethylene triperoxide diamine)

TATP (tri-acetone tri-peroxide) is the homemade explosive of choice because the materials can be easily purchased from a beauty supply shop. Almost all of the bombings in the United Kingdom and other European countries have been perpetrated using TATP, and there have been successes and failures using this concoction. One such success involved the London suicide bombings of July 7, 2005. The bombings killed 56 (including the bombers) and injured 500. The terrorists targeted London's transportation system, including:

- Circle line (b/w Liverpool & Aldgate)
- Circle line (near Edgeware road)
- Piccadilly line (between Kings Cross & Russell Square)
- Double-deck bus near Tavistock Square

Types of Large Vehicle Borne IEDs

In general, there are three types of Vehicle Borne IED operations. All three types of operations have been used worldwide.

- Unoccupied vehicle parked and loaded with explosives to be detonated, usually by remote control, when the target vehicle approaches (aka roadside bomb)
- Unoccupied vehicle packed with large amount of explosives that is parked at a target building or location to be detonated by timing mechanism or remote control. (e.g., Oklahoma City)
- Vehicle loaded with a large amount of explosives to be detonated by its driver in a suicide operation on reaching the target destination. (e.g., 1983 Beirut Embassy bombing)

Suicide Bomber Operations

An operational methodology, in which the success of an attack is dependent on the death of the attacker, suicide terrorism is the readiness to sacrifice one's life in the process of destroying or attempting to destroy a target to advance a goal. The aim of the psychologically and physically war-trained terrorist is simply to die while destroying the enemy target. Suicidal terrorist attacks can and have come in a variety of forms, including airline hijackings, personal assaults with guns or edged weapons, and improvised explosive devices or bombs. To date, however, it is the bomb that has proven the most efficient and effective method to terrorize the intended victim population.

There are basically two types of suicide bombs, differentiated by their means of concealment and conveyance to the target:

+ The hand-carried or body bomb
+ The vehicular bomb

The threat of suicide attacks is spreading. During the early 1980s, only three countries were affected: Lebanon, Kuwait and Sri Lanka (see map next page).

Since those early days, however, suicide bombings have increased dramatically. From 1982 through 2015, for example, 4,933 such attacks occurred worldwide – though more than half occurred in Iraq and Afghanistan since 9/11. Al-Qaeda and its affiliates remain the primary employers of this technique – based in no small way on Osama bin Laden's 1998 Fatwa stating that, "It is the duty of all Muslims to kill all Americans."

No country is immune. Suicide bombings will continue because they remain effective tools for terrorist organizations. The motivation for these groups, and subsequent groups, to use suicide bombs generally revolves around two beliefs:

- Religious fanaticism, where suicide bombing is seen as serving Allah through martyrdom and acquiring a desirable afterlife
- Political extremism, where suicide attack is seen as the most effective political and/or military tool (usually nationalist or leftist in nature)

Two examples of suicide bombing incidents include:

The Khost bombing, December 30, 2009: Humam Khalil Abu-Mulal al-Balawi, a Jordanian physician, was also a triple agent who worked for the Jordanian Intelligence Service and whom the United States wanted to recruit as an access agent to Osama bin Laden. Unknown to the Jordanians and U.S. intelligence, Balawi also worked for al-Qaeda.

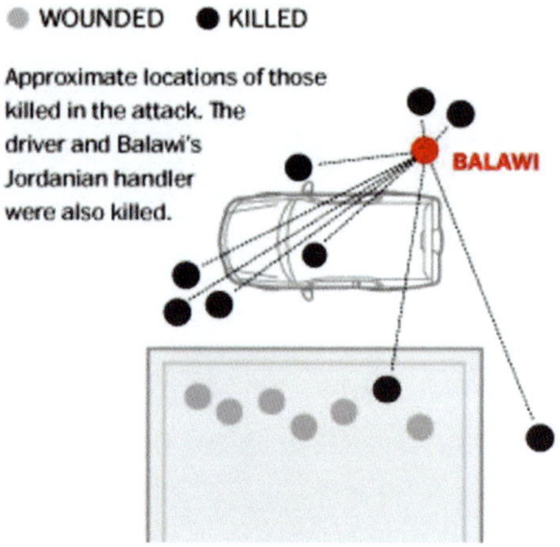

The hotel bombings in Amman, Jordan, November 9, 2005: Three hotels were simultaneously attacked in Amman by bombers using suicide vests or belts.

- Days Inn: outside hotel, three killed
- Grand Hyatt: in the hotel bar, seven killed
- Radisson: in ballroom, 38 killed

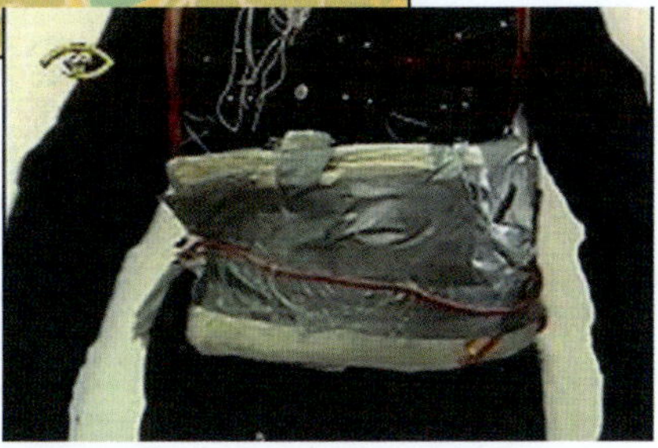

The Radisson Hotel bombing was the most significant because of the large number of fatalities. This bombing was conducted by a husband-wife team from Iraq. The wife's explosive belt malfunctioned as she tried to activate the detonator. The husband pushed her away and jumped on the table and detonated his belt. The wife was executed February 2015 in Jordan for the crime.

Note: In the weeks following the three suicide bomber attacks in Amman, Luke Bencie traveled to Jordan as part of a U.S. Government response team assisting with the investigation.

Suicide Vehicle Bombings

The 1980s and '90s saw the beginnings of suicide vehicle bombings around the world. Also included was the USS Cole suicide bombing, in which the vehicle was a small boat that pulled alongside the warship. Other such incidents include:

- U.S. Embassy, Beirut, Lebanon
- U.S. Embassy Annex, Beirut
- U.S. Marine barracks, Beirut
- U.S. Embassy, Dar Es Salaam, Tanzania
- U.S. Embassy, Nairobi, Kenya
- U.S. Consulate, Karachi, Pakistan

These types of bombings will continue, but they are based on the acquisition of large amounts of explosives, which are being controlled ever more tightly around the world.

Non-Suicide Vehicle Bombs

There have been and will continue to be non-suicide vehicle bombs used throughout the world, as long as that tactic fits the terrorists' operational objectives. Some significant incidents over the past three decades include:

- 1993 World Trade Center Bombing, New York City
- Oklahoma City Bombing
- Royal National Guard Headquarters, Riyadh, Saudi Arabia
- U.S. Air Force Khobar Towers, Dhahran, Saudi Arabia

In summary, terrorist bombings likely will be with us for a long time. The goal is to protect our infrastructure against this type of operation, which can be done. If our infrastructure is protected to such a point that the risk to the terrorist operation is great, an adversary might move on to a less-challenging target.

4. Target Analysis & Vulnerability Assessment

Hope is an expensive commodity.
It makes better sense to be prepared.
—Thucydides

The Vulnerability-Assessment Process

The planning, execution and magnitude of the catastrophic events of 9/11 were a milestone in terrorism. They exceeded pre-9/11 attacks anywhere in the world. The result is that the threat today is more of the same. Terrorist tactics will continue to be more aggressive, horrific and deadly. Simply put, nothing is immune from the most brutal acts imaginable in this day and age.

Some examples of potential targets worldwide:

- Ports
- Air and ground transportation
- Water supplies
- Food supplies
- Shipping
- Electric power
- Public or government buildings
- Sports complexes
- Entertainment centers
- Hotels
- Shopping malls
- Gatherings, groups and individuals

Given the vast array of targets of opportunity, we need comprehensive vulnerability assessments that effectively address all these threats. In addition, based on the vulnerability assessment, security enhancements must provide adequate protection to both soft and hard targets.

Security can be complex, but we need to answer the following questions:

- Is the security solution worth the additional risk that might result from the vulnerability assessment?
- Are there trade-offs that must be made, and are they worth it?

Vulnerability-assessment methodologies can take many different shapes and forms. Examples of vulnerability-assessment methodologies include:

- CCPS-SVA: Center for Chemical Process Safety-Security Vulnerability Assessment
- NRWA: National Rural Water Association Self-Assessment Guide
- RAM-D: Risk Assessment Methodology for Dams
- RAM-W: Risk Assessment Methodology for Water
- RAM-C: Risk Assessment Methodology for Communities
- VSAT: Computer-based Self-Assessment Methodology

I, for one, prefer the CARVER methodology because it is the best and most accurate way to characterize assets by determining risk and probability of attack. The following five steps support the CARVER vulnerability-assessment process:

Step One: What assets are you trying to protect? Many people never ask this question (create an asset inventory)

Step Two: What are the risks to these assets? What and who are we defending against? (design basis threat)

Step Three: How well does the security solution mitigate those risks? Does it work? What are the failure points? (countermeasures)

Step Four: What other risks does the security solution cause? Ripple effects and unintended consequences (risks to other assets)

Step Five: What costs and trade-offs does the security solution impose or create? Inconvenience, privacy, slow-down of commerce, etc. (cost benefit analysis)

Never forget that a vulnerability assessment is a process with a beginning and an end. The process, as defined in the following paragraphs, requires a step-by-step approach to the vulnerability-assessment report. After conducting hundreds of vulnerability assessments on all manner of infrastructure, I've never deviated from this step-by-step approach. As you can see, the process resembles many other vulnerability-assessment methodologies; with the exception that this methodology relies heavily on CARVER to deliver actionable findings, on which to base countermeasures and other security recommendations.

The **Planning Phase** of this process is generally done prior to the site visit. In this phase, the assessment team, which comprises security and emergency management professionals, determines what additional personnel will be needed to conduct the vulnerability assessment. Based on the infrastructure's acceptance of the vulnerability-assessment proposal, the scope and objectives of the assessment are defined.

The **Data-Collection Phase** begins when all assets have been identified and design basis threats have been created for each of those assets. Only then can security-system effectiveness for each of those assets be defined. To me, this is the fun part of any vulnerability assessment, because you put yourself into the mindset of the adversary and create plausible scenarios to compromise that asset. Obviously, this must be done onsite. When you create those plausible scenarios, you then conduct an analysis of the consequences of loss. We generally do this with the help and guidance of the facility management and their operations and engineering personnel.

In the **Assessment Phase**, we analyze all information collected in the Data-Collection Phase to determine the risk and probability of attack. This is where we apply CARVER, and it represents the final step before writing the report. As I have mentioned repeatedly, CARVER stands for Criticality, Accessibility, Recoverability, Vulnerability, Effect and Recognizability.

Completing the CARVER matrix for these criteria is a team effort. As an example, Criticality, Recoverability and Effect generally require collaboration between the security team members and infrastructure management and their operations and engineering personnel. It is unlikely that security personnel alone would have the knowledge necessary to quantify those criteria accurately.

Meanwhile, Accessibility, Vulnerability and Recognizability are generally quantified by the security members of the team.

- Criticality: Security team members and infrastructure team members
- Accessibility: Security team members
- Recoverability: Infrastructure team members
- Vulnerability: Security team members
- Effect: Security team members and infrastructure team members
- Recognizability: Security team members

For example, when I conducted my first water-vulnerability assessment, I knew nothing about water infrastructure. This happens a lot. When you enter a new infrastructure that you haven't had the opportunity to assess, it immediately becomes a vast learning experience. In that first water assessment, I joined myself at the hip to the engineering and operations personnel of the utility to learn as much as I could. My goal was to keep them operating, and I needed to know what was critical and what the effect would be if they could not recover quickly from a loss. I obtained my education from utility management briefings, engineering briefings and operational briefings prior to assessing the assets.

The **Reporting Phase** consists of compiling all the data accumulated in the prior phases of the vulnerability-assessment process. This is the most important phase, because this is the product you leave behind after you and your company ride off into the sunset. This is how you will be remembered, and this is what will follow you from assessment to assessment. A new client will want some indication of past performance and usually wants to contact your previous clients to determine your proficiency. In many cases, this comes right from the report you leave behind and how well that report was received by your client.

I won't go into detail on exactly what's in the report, but I can give you the elements/content that must be considered.

Executive Summary

The Executive Summary provides the reader with an abbreviated view of the VA Report.

It contains information regarding the facility(s), operation(s) or program(s) that were identified as assets, the threat to those assets, the prioritization of those assets and the level of risk identified for each asset.

It also includes the factors involved in setting up the team, a summary of the information collected in the assessment phase, and the final report summary of risk-management strategies suggested by the VA team. Its essential elements include:

- Introduction
- Objectives
- Description of facility
- Description of operations
- Assessment team
- Critical assets (top 10 CARVER chart)
- Threats (threat spectrum)
- Effect of loss
- Risk reduction/countermeasures

Introduction, Scope and Objectives

The goals we are trying to achieve in this section encompass a number of elements, but most important is the introduction of you and your company to establish credibility in conducting a vulnerability assessment. The scope of the assessment is described in detail, including the planning process, which will lead to a successful assessment. The objectives are derived from either the mandates for the vulnerability assessment or the client's requirements. Its essential elements include:

- Introduction of assessors
- Introduction of infrastructure
- Purpose for the VA
- Scope and objectives
- Team selection

VA Methodology Application – This section provides a description of the methodology used to conduct the vulnerability assessment. Its essential elements include:

- Introduction
- Assessment criteria used

Threat Assessment

The VA team evaluates the external threats, the internal threats and the combination of both. The adversaries are identified and characterized. All possible and reasonable local, regional or even national intelligence information should be included in the threat assessment to substantiate the threat. Design basis threats are included as a roadmap for the vulnerability assessment, including these essential elements:

- Introduction
- Threat assessment (international, domestic and insider)
- Adversary capabilities
- Threat spectrum

Asset Characterization/Consequence Assessment

This section identifies and describes the assets and their operational importance. (An asset can be a facility, operation or program.) Once again, the CARVER criteria used to characterize the assets are Criticality, Recoverability, Vulnerability, Effect and Recognizability, and the essential elements include:

- Introduction
- Exterior and interior assets
- Threat to assets
- Consequence of loss

Security System Effectiveness

A description of a system, sub-system, complex, asset or operation to determine its ability to effectively deter, detect, delay and respond to an attack. The CARVER criterium used to assess SSE is Accessibility, and its essential elements include:

- Introduction
- Identifying assets

- Identifying paths of likely success
- Identifying potential scenarios
- Identifying existing countermeasures and level of effectiveness
- Identifying operational vulnerabilities

Physical Security Implementation Plan

The security implementation plan outlines security upgrades, modifications of operational procedures and/or policy changes to mitigate identified risks to critical assets. The plan serves as a basis for comparing the cost of protection against a perceived risk. Its essential elements include:

- Access procedures
- Community security awareness
- Emergency response plan
- Background investigations
- Information protection
- Key management
- Photography
- Public information requests
- Response force
- Suspicious incident reporting
- Threat detection training
- Threat incident plan
- Tiered physical security response plan
- Tours of facilities
- Workplace violence program
- Employee security awareness training
- Security equipment

> NOTE: During The CARVER Target Analysis and Vulnerability Assessment Training Course ™, offered by Security Management International, information is provided that goes into detail on writing a complete report that experience has shown is accepted by all manner of clients.
> Detailed information on the CARVER course can be found at:
> https://smicounsultancy.com/the-CARVER-methodology.html

Target Analysis Process

A target analysis is an offensive methodology used to identify a target that, if compromised, meets a prescribed strategic or tactical objective. Even the term "Target" is used offensively to define a person, environment, facility, material, information, piece of equipment, reputation or activity that, if destroyed, meets a strategic or tactical goal.

Vulnerability assessment, meanwhile, is a defensive methodology. We use it to determine the likelihood that an adversary can successfully exploit a system's or asset's vulnerabilities. An asset is any person, environment, facility, material, information, piece of equipment, reputation or activity that has positive value. That is the key to this definition. If there is no positive value, there is no need for potentially expensive countermeasures.

Both definitions represent the basis for conducting an effective vulnerability assessment, because we use both the target-analysis and vulnerability-assessment processes in the analysis. Simply put, we conduct both the target analysis and vulnerability assessment simultaneously.

During the analysis of the site's Security System Effectiveness, the assessor puts himself into the shoes of the adversary and tries to determine plausible scenarios to compromise the target or targets. Meanwhile, we also try to determine countermeasures to prevent a potential scenario from taking place that would compromise the asset.

Any worthy adversary will conduct both a target analysis and operational plan to attack any given target. They might not call it a target analysis or an operational plan, and the process might proceed subconsciously, but it will be done.

It's appropriate at this point to define the term "adversary," because an adversary is not always a terrorist. An adversary could be any individual or group, organization or government that conducts activities, or has the intention and capability to conduct activities detrimental to an asset. Adversaries might include intelligence services, political activists, terrorist groups, criminals, rogue employees and/or other private interests.

The target analysis collects and examines the available information required to meet – and counter – the strategic or tactical objectives of an adversary that meets his presumed requirements irrespective of his capabilities. This was born out by the 9/11 attacks, in which some of the 20 terrorists involved needed to learn to fly large commercial aircraft. They were not particularly concerned with landing and takeoff, but they did need to learn how to aim an airplane at a target.

This was a complex operation, years in the making. But now this type of operation is probably no longer feasible, because much of the world has enacted countermeasures against 9/11-style attacks – though terrorists continue to commandeer aircraft sporadically.

More recently, simpler operations have become common, such as random stabbings and using trucks and other vehicles to mow down defenseless pedestrians. Such acts, though horrible in their results, require little planning, logistical support or funding.

Adversaries conduct analyses against targets of interest that, they believe, will increase their chances of success. Such analyses tend to proceed as follows:

Strategic = System + Subsystem

Tactical = Complex + Asset/Target

Here is a more detailed example of the analyses a potential terrorist might conduct in planning to achieve a hazardous material release:

EXAMPLE: Terrorist Perspective HAZMAT Release

Systems: Water
Communiications
Power
Transportation

Subsystem: Source
Disinfection
Storage
Distribution

Complex: Water Treatment Plant
Reservoir
Pumping Station

Asset / Target: Chlorine Storage/Operations

The example provides a visual of how a terrorist would select a target for a hazardous-material-release operation using the target-analysis process. What's particularly frightening is that all the information required to conduct this operation can readily be found on the Internet.

The Target Analysis Methodology

The targeting process comprises three phases:

1. Statement of Requirement. Senior mission planners develop the SOR, which provides the basis for beginning the target analysis. It also defines the objectives and goals to be met and lists the requirements to be fulfilled.

2. Target Analysis. The objective is to select a target that meets prescribed requirements identified by the SOR.

3. Operational Plan. This is where the rubber meets the road. The mission planners determine Who will conduct the operation, What assets will be required, Where the operations will take place and How the attack scenario will unfold.

Based on the information acquired during the analysis process, planners create a target folder. This is a composite study of pertinent material relating to a particular target or target complex.

Folders are created based on the prioritized results of the CARVER matrix. Their purpose is to provide senior planners with all known target information.

When the team develops a target analysis of an infrastructure by the military and/or intelligence services, they will use the process provided above, which has not changed and has been refined and perfected over the years. They use CARVER to identify the target or asset to be attacked based on their Statement of Requirement.

The CARVER criteria – Criticality, Accessibility, Recoverability, Vulnerability, Effect and Recognizability – are evaluated on a scale of 1 to 5, and target attractiveness is determined based on the sum of those values. The higher the number, the greater the attractiveness of the target.

Target Analysis Conclusions

Even though target analyses and vulnerability assessments use the same methodology, there are differences between them. In a vulnerability assessment, you receive the complete support of the infrastructure's team to gather the information necessary to secure the asset. target analysis, however, does not afford such a luxury.

When we conduct target analyses, we offer the following guidelines to the personnel at the facility in question:

- A target analysis will be conducted by your adversary against your asset prior to an attack.
- The target analysis will be based on your adversary's capabilities.
- A target analysis is going to be done without your knowledge.
- A target analysis is going to be done without your knowing support.
- Your adversary's capabilities will determine your vulnerabilities.

This is why it's so important to conduct both a target analysis and a vulnerability assessment in parallel. You must determine and understand what knowledge the adversary needs and/or can obtain on your infrastructure without your help.

5. Security System Effectiveness

Without data, you're just another person with an opinion.
— Edwards Deming

HERE, IN THE ONSITE PHASE of the vulnerability-assessment process, we evaluate a system, subsystem, complex, asset or operation to determine its ability to, effectively, deter, detect, delay and respond to an attack.

In short, this is where we walk the walk and talk the talk. We attempt to identify adversary strategies and paths of least resistance by creating scenarios directed against each critical asset, and we document Security System Effectiveness for all adversary types and undesired events.

We do this by identifying:

- Target assets
- Paths of least resistance
- Potential adversary scenarios
- Operational vulnerabilities
- Existing countermeasures and their effectiveness

The elements of Security System Effectiveness are deterrence, detection, delay, response; and mitigation. Let's look at each one of these elements separately to define what it can accomplish as part of the site assessment.

Deterrence. This is a countermeasure strategy intended to discourage a breach of security by means of fear or doubt. The countermeasures can be real or imaginary. Deterrence is probably the best defense against an insider threat, which, other than an active shooter, is usually limited to theft, vandalism, sabotage, etc. In such cases, the insider generally does not want to get caught, whereas an active shooter generally isn't concerned with getting caught, believing he or she will not survive the encounter. Some examples of deterrence are motion-activated lighting, penalties, CCTV, no-trespassing signs and other real or imagined countermeasures.

Deterrence played a big part in the early and mid-1970s, when international terrorists began targeting embassies, airlines and private businesses, among other venues. Increasing security at those sites generally encouraged terrorists to consider other targets that offered more chances for success. In other words, don't attack my embassy; attack the one down the street with weaker security.

Detection	Delay	Response
Intrusion Sensing	Passive Barriers	Interruptions (comms/deployment)
Alarm Communication	Active Barriers	Neutralization
Alarm Assessment		

Detection, Delay and Response

Detection is a countermeasure intended to identify inappropriate, incorrect, malicious or anomalous activity by an adversary attempting to breach security. This can be accomplished using cameras with motion sensors, passive IR, microwave sensors, door alarms, etc. The most important component of detection, however, is the ability to assess an alarm once received. All alarms and security sensors are vulnerable to false activation. Too many false alarms suggest the sensor should be desensitized, an inherent weakness in such systems.

More recently, managers of much of the critical infrastructure throughout the world have installed fixed CCTV systems with motion-sensor capability. These systems not only detect potential intruders but also provide immediate assessments of the cause of the alarm. Pan/tilt/zoom cameras provide enhanced assessments and backup for the fixed CCTV installations.

Detection criteria should assure that any person crossing the security perimeter will be detected, whether walking, running, jumping, crawling or climbing at any point inside the detection zone. It should also provide the earliest possible detection of unauthorized intrusion as far away as possible from the security subject.

Delay is a countermeasure strategy. It's meant to slow the progress of an adversary, usually via various barriers, until security can mount an adequate response. Examples of delay strategies include locking devices, fences, doors, walls, etc.

Delay can be crucial to security system effectiveness, assuming an asset can be protected. From threat- and vulnerability-assessment standpoints, it is critical to determine how much time is available between detecting the intruder and the intruder reaching the asset. Then you must determine if the delay is sufficient based on the time the response team requires.

For example, if an adversary's task time is 10 minutes, and the response time is 20 minutes, then you need a delay of at least 10 minutes. If the response time is greater than the adversary's task time, and a delaying tactic isn't available, then it becomes plausible that an asset cannot be secured or protected. That situation requires mitigation plans to deal with a lost asset.

Response is a security strategy that requires reacting to an adversarial activity after detection and assessment. It can include notifying authorities, evasive action, confrontation and apprehension. Some examples of response include onsite security, local police and/

or employees. But these types of responses are generally inefficient, because their effectiveness depends on the time it takes to interdict an adversary, and in many cases the time available is insufficient. Even when alarm mechanisms are located at the asset – such as magnetic switches on doors providing access to the asset – usually when the alarm sounds the adversary has already reached the asset, ready to do his dastardly deed. So, an alarm, even if received by the response force, won't always provide an opportunity to interdict the adversary.

Some responses involve technology rather than people. These

include automatic system shutdowns and/or automatic locks. In the case of a water utility with numerous fresh-water reservoirs servicing a large sector of the population, the reservoirs could be contaminated from an outside source. Uncontained contamination of a reservoir can spread throughout the water-distribution system, possibly requiring the entire system to be flushed and cleaned, thereby shutting down the whole water supply. Automatic system shutdown is one way of reducing the effects of contamination: If the hatch is opened, the reservoir is disabled and will not allow water to flow into the distribution system.

Mitigation reduces the severity of a consequence. Each infrastructure contains assets at risk from natural and manmade hazards. You identify those assets broadly to include everything important to the function of the infrastructure. Although all assets might be affected by hazards, some are more valuable because of their physical characteristics. The purpose of an asset inventory is to identify specific vulnerable assets in your infrastructure.

When compiling a mitigation plan, the planning team analyzes the asset inventory to reflect current threats. Some examples of mitigation include:

- Redundancy
- Backup systems
- Spare parts
- Interdependencies
- Excess storage
- Split critical assets
- Repair capability

Each of these examples would be incorporated into the mitigation plan if applicable.

Perimeter and Building Security

As I outlined in the beginning of this chapter, Security System Effectiveness is a key element of the vulnerability-assessment process. I break down vulnerability assessment into two key elements: the perimeter and what is located inside the perimeter, which could be a generator, transformer or other asset of the infrastructure, or it could be a building without a traditional perimeter. I handle assessments of a perimeter and a building in much the same way, starting on the outside and working toward the interior.

Perimeter. The assessment should always start at the perimeter. This is the outermost boundary of a property-protection system. It can support access control and detection capability. Its primary function is to exclude unauthorized personnel from gaining access

to your asset. It also establishes a visible legal boundary demonstrating your intent of denying an intruder access. It also supports signage such as, "No Trespassing," "Electronic Surveillance in Use" and/or "Use of Deadly Force Authorized," all of which should act as deterrents to would-be intruders.

One of the key elements of perimeter security is the fence. There are many types of fences, and the fabric of the fence will determine the security level. For example, the smaller the mesh the more difficult to climb; the heavier the gauge more difficult to cut. The common chain-link fence is the least-secure fence construction

because it can be bypassed relatively easily. There are fixes for chain-link fences, however, which can protect against high-speed (50 mph) vehicle penetration of your perimeter. The fix involves using 1-inch steel cable at a height of 2 to 3 feet and anchored every 100 feet. In some cases, two cables are used to prevent large-vehicle penetration of the fence line.

As mentioned, the perimeter is your starting point, so what are you looking for? You're looking for breaches or potential breaches in your perimeter such as washouts at the bottom of the fence fabric, bridges such as tree limbs or equipment positioned against or near the fence, too much vegetation, and tunnels such as culverts, drains, creeks and debris. You're also looking for removable objects such as rocks and gravel, and you want to make sure the fence has signage that's connected properly and legible from the outside. You need to pay attention to the access-control points such as gates. Are gates properly locked, for instance? Are gates numbered and consistent with the design of the fence line?

Most important, what do I see from outside the perimeter? Remember that the adversary will begin casing your assets from there. Outside the perimeter is considered a safe space for him or her. Following are some examples of what you might be looking for from outside the perimeter.

- Do I see potential targets?
- Can I recognize security systems?
- Can I recognize the path of least resistance?
- Can I develop an attack scenario?

In addition, can I find signs of recent activity or tampering in places where there should be no signs? This could include tire tracks outside the perimeter, footprints, litter such as cigarette butts at potential observation sites and evidence of tampering with bolts, nuts and connections.

Case Study: International Convention Center (ICC)

The following incorporates all of the elements involved in the security-system-effectiveness assessment. It involves a thorough examination of a facility's perimeter and the building complex located inside that perimeter. The assessment covers an international convention center somewhere in Asia.

The center hosts many functions and events throughout the year. Many are high-profile international events such as the Annual Asian Ministers Meeting. Many of the attendees are heads of state, including the President of the United States, other VIPs, staff and conference attendees, all of which makes the center an inviting target for adversaries.

The vulnerability assessment took approximately seven days to complete because of the internal and external complexity of the infrastructure. Our preferred assessment tool: CARVER. In this case, however, we used a combination of Sandia National Laboratory's Risk Assessment Methodology (RAM) supplemented with CARVER. Both methodologies assured that the recommended countermeasures would address the facility's vulnerabilities.

As mentioned earlier, CARVER can be used to augment many other vulnerability-assessment methodologies, and that was the case here. The assessment team wanted granularity in asset prioritization to apply cost-effective countermeasures. CARVER and RAM easily accommodated the physical and operational challenges. We applied the methodologies in a logical, stepped fashion that considered the threat, target identification, types of aggressors, malevolent acts of concern and consequences of a successful malevolent act.

The threats considered in this vulnerability assessment included:

- Vehicle borne IEDs
- Placed IEDs
- Body-worn IEDs
- Small arms and assault weapons
- Chemical, biological and radiological attacks
- Theft and Vandalism

Although we considered theft and vandalism, they were not our primary concerns because the effects of each would not be catastrophic to the structure or an event.

Facility Description

The convention center is a government building under the management of the country's Ministry of Foreign Affairs. It is located close to the city center and surrounded by a number of other venues such as a sports stadium. The building comprises a plenary hall, multi-purpose hall, a royal lounge, news conference hall, entrance concourse, summit rooms, four bilateral rooms and six pre-meeting rooms.

Each venue posed unique security issues attributable to the wide variety of functions and attendees. The vulnerability-assessment team addressed each venue separately because the risks to each were substantially different. As you might imagine, this was an enormous undertaking. Each of the venues or rooms required separate assessments.

For example, the center has a defined perimeter composed of chain-link fencing with triple strands of barbed wire and chain-link fencing without barbed wire covering 75 percent of the property perimeter, as well as decorative, wrought-iron fencing covering the remaining 25 percent.

The perimeter was extensive, making security and access control problematic. The team found many areas where the perimeter could be breached without detection. Based on this premise, the team concentrated on the main building, using its external walls as the functional perimeter. This might seem to go against my recommended procedure, in which the perimeter is used as the starting place to conduct the vulnerability assessment. But after studying the perimeter, we determined that the most important asset was the main building, which had to be assessed even more carefully for vulnerabilities followed by crafting the

countermeasures needed to reduce or eliminate those vulnerabilities.

Our team had to be versatile and adjust our assessment to the asset, taking into consideration its uniqueness.

Access-Control Points: Perimeter Gates, Locks and Vehicle Barriers Perimeter Gates and Locks

Access-control points are just that: facility components that define authorized access to personnel and vehicles. Simply put, access-control points allow only authorized personnel and vehicles, and they restrict unauthorized personnel and vehicles.

Most access-control points, however, offer limited security because many of them are unmanned and do not operate properly. Take vehicle gates. There are two common types: swing gates and sliding gates. Most activate slowly, thereby allowing piggybacking or high-speed entry.

I once conducted a vulnerability assessment of a chlorine repackaging facility that held railroad tank cars filled with chlorine. The vehicle gate leading into the facility was a sliding gate, which

operated slowly and could allow multiple vehicles to enter. Because the facility was adjacent to high-rise condos, release of a large quantity of chlorine could be catastrophic. So, I recommended drastic measures to protect against a vehicle bomb attack. The solution was to install double gates like the sally ports that most prisons use. One gate opens and lets a vehicle in, but the vehicle is stopped by the second gate, which cannot open until the first gate closes. This was a very expensive fix, but the consequences of an attack would be so great that the extra expense was warranted.

Pedestrian gates vary greatly in their security level and ability to prevent unauthorized access. Security in many cases depends on the locking devices used to secure the gate. The devices can be electrically operated by a pin code and/or magnetic card, or they can use mechanical devices such as padlocks or other types of key-activated locks. One type of pedestrian gate is a turnstile, which is mainly used in high-security facilities. The turnstile is activated by a magnetic card or pin code, and it allows only one person to enter at a time. The person entering has only a few seconds after deactivating the lock to go through the turnstile, or else that individual will be locked out until security can respond to verify credentials. Another type of gate-locking mechanism is the use of cipher code, which deactivates a deadbolt or releases a magnetic lock. Such mechanisms tend to wear over time and can grow progressively weaker.

Key-locking devices on gates are the least secure. The lock can be bypassed by cutting or brute force. In many cases, facilities leave gates open and unlocked for ease of access throughout the workday. Another problem is lost or misplaced keys. Even key-hierarchy management is a widespread problem because in many cases companies lose track of who holds keys. This happens, for example, when employees leave a company and their keys are not collected.

A major problem with padlocks is the propensity to use them with chains. In such cases it doesn't matter how good the padlock is; the chain becomes the weakest link (no pun intended). In instances where multiple entities require access to a compound through the same access points, multiple locks are used in the form of a daisy chain. In other words, each lock is locked to an adjacent lock forming a chain in which any lock can compromise that asset. This further degrades security because any missing key also compromises

the asset. This type of configuration is generally found at the gate entrance to elevated above-ground reservoirs. At these sites, the water utility would need access as well as cell phone companies and 911 maintenance personnel, etc.

Vehicle Barriers

Vehicle barriers, which can be active or passive, are used to control and/or prevent vehicles from entering the perimeter or building. There are many types of physical barriers such as static or retractable bollards, pop-up barriers, drop-down crash-rated arms, sliding gates, cable-reinforced chain-link fence, etc. In addition to restricting vehicle access to infrastructure assets, the barrier should also provide a safe standoff distance from critical assets or buildings. As in the case of a vehicle-borne improvised explosive device (VBIED) the perceived threat and size of the vehicle will determine a safe standoff distance. However, in most urban situations obtaining a safe standoff distance can be difficult at best. Some of the most common protection devices used to prevent vehicle penetration are crash-rated bollards installed as far as possible away from the asset or building. Bollards can be active or passive, but placement is critical. They must be spaced correctly to prevent a vehicle from slipping between them or going around them. During many vulnerability assessments, I have seen bollards placed incorrectly, because the process wasn't supervised by competent security personnel.

Buildings

What is a building? This might sound like a simple question, but it isn't. There are many types of buildings, and each might have a distinctive purpose that could make it more or less attractive to an attacker. There are office buildings, both

commercial and government; there are hotels, schools, malls, sports venues, transportation hubs, churches and many other types. Most, if not all, are occupied by staff and visitors throughout the day and night, which increases their attractiveness to some adversaries. Most have little or no security against even known threats. Having lived in the Washington, D.C., area, I have visited a lot of government buildings. I must say they take security seriously. When I've visited nongovernment buildings, however, I've seen much less security. In many instances, there might be a sign on the entrance door that reads "No Guns Allowed." But as we know, almost every active shooter incident has occurred in a gun-free zone. Consider the old saying: "When seconds count, police are minutes away." Whenever personal or personnel security is concerned, management must deal with this threat – as well as accommodating jurisdictions that allow concealed carry.

When I assess Security System Effectiveness, I start with the exterior of the building, particularly when an outside perimeter (fence) does not exist. I conduct the assessment just as I would conduct a perimeter assessment. In other words, I treat the exterior walls as the perimeter; the doors and windows as access points, and I begin my assessment there, working toward the inside. I also consider nontraditional access points, such as underground grates, rooftop access, etc.

As I asked in the previous section on perimeters, what are you looking for? In the case of a building, you look for the same things as when looking at a traditional perimeter.

- Are you able to define surveillance locations?
- Are you able to recognize critical assets?
- Are you able to define paths of least resistance to a potential target?
- Are you able to identify an attack scenario?
- Are you able to identify security systems?
- Are you able to identify non-traditional access points?

Threat Spectrum

There are many and varied threats against a building and its occupants; some are more likely than others. This is a good thing, because some of the most egregious threats are less likely to happen, whereas theft and vandalism are common. As an example:

- Vehicle-borne IED attack (suicide/non-suicide)
- Body-worn IED attack (suicide)
- Placed IED
- Standoff weapon assault
- Chemical/biological contamination
- Armed assault (outsider/insider)
- Armed assault (hostage/barricade)
- Cyber attack
- Vandalism
- Theft

Vehicle-borne IED attack tops the list for good reason. This is a high penalty threat, which I need to cover in some detail. As mentioned previously, there are three types of attacks using the VBIED.

1. An unoccupied vehicle parked and loaded with explosives to be detonated, usually by remote control, when a target approaches the vehicle

2. An unoccupied vehicle packed with explosives and parked at a target building or location to be detonated by timing mechanism or remote control

3. A vehicle loaded with explosives to be detonated by its driver in a suicide operation on reaching the target destination

Let's focus on the attack scenarios described in 2 and 3, both of which directly affect buildings. Our main goal should be to prevent the VBIED from getting inside the security perimeter. But with large VBIEDs, even keeping the vehicle outside can still cause an enormous amount of damage and casualties, such as in the case of the Alfred P. Murrah Federal Building in Oklahoma City. That 1995 terror incident involved approximately 4,000 pounds of ammonium nitrate and fuel oil, and it detonated directly adjacent to the building, killing 168 people.

Many VBIED incidents involve suicide attackers attempting to get the vehicle inside a building or as near to it as possible. Examples include the truck crashing through the front door of the U.S. Embassy in Beirut, Lebanon, in 1983; the U.S. Marine Corps barracks bombing in Beirut in 1983, and the twin bombings of the U.S. embassies in Nairobi, Kenya, and Dar es Salaam, Tanzania, in 1998.

Since 1982, the world has seen many thousands of VBIED attacks, although most of these have taken place in Iraq and Afghanistan. Recently, a new threat involving vehicles has emerged. These attackers, eschewing explosives, use their vehicles as guided weapons against pedestrians and crowds, such as the recent incidents in London and in Nice, France. Many of the vehicle countermeasures we have identified earlier perhaps could have been effective against this new threat. The problem is the amount of real estate to be protected, which increases dramatically if you try to protect every venue where crowds gather.

If your aim is to reduce threats against buildings and occupants, you must incorporate access-control measures to limit access by unwanted individuals. You can install access-control hardware and systems that will significantly increase building security. Some hardware and systems currently used include biometrics such as fingerprint readers, hand scans, voice authentication, retina scans and iris scans, as well as outside lighting devices that include motion-activated and vandal-proof fixtures along with standard lighting. A variety of locks, alarms and other intangible security measures play a role in preventing unauthorized access. We covered some of the security devices above. Intangible security measures include simple practices such as displaying identification badges correctly. Other available technologies include walk-through and handheld metal detectors, package x-ray scanners and radiation detectors.

As you can see, no single piece of hardware or system will provide total security for a building. Total security must include a layered approach using a mix of hardware and systems, all combined effectively to secure the facility.

6. Utilizing CARVER as Part of a Security Management Consulting Process

Culture eats strategy for breakfast.
—Peter Drucker

BECAUSE THE CARVER METHODOLOGY PROVIDES a framework for analyzing and measuring vulnerabilities, it can be easily incorporated into a security management consulting model. Just as strategic management consulting firms such as McKinsey & Co., Bain Capital or Boston Consulting Group use statistical modeling to demonstrate their findings and recommendations, the same philosophy holds for security management firms. Some would claim that management consultants take what the client already knows and simply repackage it into fancy charts and graphs. This is partially true. The Harvard MBAs from the top five consultancies have a reputation for dissecting a client's current situation and demonstrating back to the client in a colorful slide deck.

In addition to their ability to visually present findings and conclusions to their clients, these management consultancies also have developed their own patented methodologies for dissecting problematic situations. For example, McKinsey has their famous MECE (pronounced "me see") methodology, which describes a way of collecting and organizing information that is "Mutually Exclusive, Collectively Exhaustive." According to Victor Cheng, a former McKinsey consultant and founder of the website casestudies.com, "When considering solving any business problem such as 'What are this company's options for growth?' 'How can we cut costs?' or 'How can we increase units sold?' the MECE principle suggests that all the possible causes or options to be considered in solving these problems be grouped and categorized in a particular way. Specifically, all the information should be grouped into categories where there is no overlap between categories (mutually exclusive) and all the categories added together covers all possible options (collectively exhaustive)."

Just as McKinsey might also advocate their 7-S framework (which helps facilitate organizational strategy), or Boston Consulting Group will extol on the virtues of their decades-old Growth-Share Matrix, security consultants need a clear process to collect, organize and share information with clients. This is where the CARVER methodology becomes extremely useful.

As security management consultants, the aim of our profession is to become trusted advisers to our clients and provide them with intelligent solutions to their complex security problems. Hence, our tagline at SMI: Your global partner for intelligent security solutions™.

One of our initial responsibilities is to educate our clients about the differences between a security consultant and a security contractor. Unfortunately, a stigma attaches to the word "security." The term has a very diverse meaning. As a security consultant, it is easy to get lumped into the categories of security guards, IT guys, bodyguards, alarm or camera vendors, and even mercenaries. Rather than being acknowledged as peers in the business community, security people can be negatively portrayed as so-called knuckle-draggers. Even security directors who are fully-employed by companies are frequently perceived as nothing more than "corporate cops." For these reasons, it is imperative that security consultants strive to set the record straight via client education and through demonstration of their knowledge, abilities and professionalism. CARVER provides a vehicle for security consultants to showcase these talents.

The SMI Security Management Consulting Process™

At SMI, what separates us from our competitors is not only our vast global security consulting experience but our corporate culture as an intelligence advisory firm. We reject the stereotypical label of "security company." Our business model is more attuned to that of a serious management consultancy. We are not security guards, nor do we operate in that type of professional space. At SMI, we take an entirely different corporate approach to security management. For this reason, we created our patented SMI Security Management Consulting Process™. It is what helps us walk our clients through the various phases of our advisory work.

We have structured the SMI Security Management Consulting Process™ into phases that are easy for our clients to understand. The process is based on the principles of CARVER, including performing both quantitative and qualitative analyses and assessments of a situation, as well as identifying and understanding the threats involved. Much like a traditional management consulting model, our process develops a hypothesis to improve the client's condition through intelligence gathering, assessment, integration, evaluation and training. Moreover, it provides ongoing mitigation strategies to monitor, prevent and respond to future security concerns.

Here is a breakdown of the phases of our process:

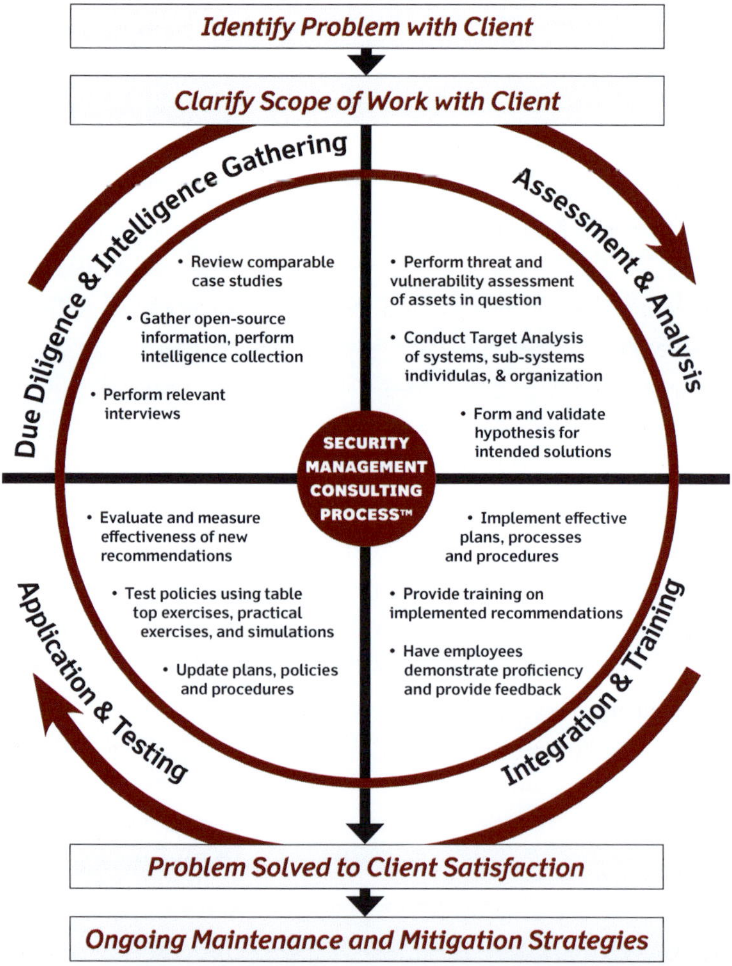

Step One: Identify the problem. The initial step in any consulting engagement is determining the problem that needs to be resolved. Much like a doctor examining a patient to diagnose an ailment, it is important to conduct a frank discussion with your client about what they believe their problem is. Many times, the client might mistakenly identify the security issue as something else. This is where the subject matter expertise of the consultant comes into play. Here again, the difference between a security consultant and a security contractor is critical. A consultant is brought in to diagnose and solve a problem. A security contractor is a hired hand brought in to perform an outlined job function. The most successful consultants know the difference and demonstrate it to their clients.

Step Two: Clarify the work with the client. When you have identified the initial problem, which will invariably change with the findings and evolution of the consulting engagement, you must clarify the scope of work with the client. By outlining the parameters of the project together, there will be no misunderstanding about roles and responsibilities. More important, this will reduce the changes of so-called scope creep, that dreaded experience of the client asking the consultant to perform more and more work that was not originally commissioned – and at no extra fee.

Step Three: Perform due-diligence and intelligence-gathering. Here is where the SMI Security Management Consulting Process™ begins to resemble elements of the CARVER methodology. Just as one would perform a threat assessment, prior to beginning a vulnerability assessment, the security consultant must also perform an initial fact-finding of the client's situation. This process includes reviewing comparable case studies, gathering open-source material, informing/performing intelligence collection and performing relevant interviews with employees, sub-contractors, vendors and/or customers.

Step Four: Perform assessment and analysis. Here, we involve the elements of the CARVER methodology. This is the meat of the consulting engagement. In this stage of the process, we perform the threat and vulnerability assessments of the assets in question. We conduct our target analysis of the systems, subsystems, individuals and organizations. We also begin to form and validate our hypotheses for intended solutions.

Step Five: Perform integration and analysis. In this phase, we begin to implement effective security plans, processes and procedures. If necessary, we provide training to the client on the implemented recommendations, and it's important that employees demonstrate their proficiency with the training. As is standard practice with all training, it is important to provide two-way feedback between client and consultant.

Step Six: Complete application and testing. As is crucial in any consulting engagement, this phase involves evaluating and measuring the effectiveness of the recommendations provided by the consultant. Unlike the stereotypes, consultants should never give a recommendation and then walk away from the project. At SMI, we see the project through to completion. As such, we will test policies using tabletop exercises, practical exercises and simulations. We will also update plans, policies and procedures following the outcomes of those exercises.

Step Seven: Solve the problem to the client's satisfaction. We consider this the debriefing phase with our clients. We sit down together and review the outcome of the engagement and ask the client, "Was the problem solved to your satisfaction?" If not, we must do everything possible to rectify the situation.

Step Eight: Conduct ongoing maintenance and mitigation strategies. It is rare that a security consulting engagement is considered a one-and-done event. Security threats are constantly evolving, and it's important to maintain an ongoing relationship with clients to keep them apprised of developments in the security environment. This could mean a retainer or agreement to conduct periodic security reviews or checkups.

Why CARVER Works

Clients often exhibit limited attention spans – and budgets – for security-related matters. Because security departments are non-revenue-generating entities for businesses, they're usually considered liabilities, or expenses, on most company balance sheets. Dedicating more money for an outside consultant to deal with security problems is usually the last thing to which a CFO wants to allocate precious revenue. Therefore, it is imperative for a security consultant to justify his/her value in both problem-solving skills and monetary terms.

Management consultants have been overcoming such obstacles for years. As illustrated above, they have demonstrated their value by delivering their findings through process visuals such as the 7-S Framework or the Growth-Share Matrix. Unless you are an engineer or scientist, most people in the business world prefer to take in information using the simplest-to-understand format possible. Fortunately, the CARVER matrix, with its six categories ranked on a scale from 1-5 (or 1-10) enables clients to understand what could be a complex security issue. Couple this with CARVER's longstanding reputation as a valuable target-analysis and vulnerability-assessment methodology, and the security consultant now has another impressive tool to add to his/her consulting toolbox.

Summary

In addition to its numerous other attributes, the CARVER methodology serves as an effective security management consulting process. Keep in mind, however, that although presenting a deliverable report utilizing the CARVER numeric scoring system is important, the consultant should never lose sight of a client's final objective: finding an effective solution. It is easy to get hung up on pretty charts and graphs, but never forget that the aim is to solve the underlying security problem. Therefore, strive to improve their overall condition. This will make you a trusted adviser and encourage repeat business.

7. Utilizing CARVER as Part of Targeted Intelligence Collection Campaign

There is one evil that I dread, and that is their spies.
—George Washington on British Intelligence, 1776

Introduction to Intelligence

The concept of intelligence – or intelligence-gathering operations – is the process by which governments and other organizations systematically collect, evaluate and exploit sensitive information for the purposes of discovering the capabilities and intentions of their rivals or enemies. It is a complex system, with the underlying objective to assist politicians, military officers or business leaders in making better-informed decisions. As nations or businesses attempt to safeguard their information from opposing collectors, the intelligence can often be inaccurate or incomplete, leaving intelligence analysts able only to provide their best estimates to policymakers. The concept of intelligence-gathering has jokingly been referred to as the world's second oldest profession. The mission statement of the Central Intelligence Agency further outlines the purpose of intelligence-gathering:

> *Preempt threats and further the U.S. national security objectives by collecting intelligence that matters, producing objective all-source analysis, conducting covert action, as directed by the president, and safeguarding the secrets that help keep our nation safe.*

In many instances, a statement of requirement will come down from the top decision-makers (though for the CIA that requirement comes from the President), requesting as much useful intelligence as the core collectors (those operations officers who physically obtain the information in the field) can acquire. When this specific task is ordered, a five-step process known as the Intelligence Cycle is implemented. This cycle is a time-tested system of checks and balances, including:

- Planning and direction
- Collection
- Processing
- Analysis and production
- Dissemination

CARVER was initially developed by intelligence officers from the Special Devices Branch of the CIA (a branch of the organization originally tasked with counterterrorism and covert targeting responsibilities). So, it only makes sense that the methodology would be used to identify potential intelligence sources as well as agents of influence. For this reason, CARVER has been instrumental in complementing elements of the intelligence cycle, as well as helping core-collectors and analysts fulfill their objectives for decades.

CARVER as Part of the Recruitment Cycle

In the parlance of the intelligence community, the process of cultivating someone to reveal information – i.e., spy or commit treason against their own country or organization – is called the recruitment cycle. A case officer, or operations officer/core-collector, is the title given the person responsible for clandestinely collecting this foreign intelligence. To accomplish his/her mission, the case officer primarily targets individuals through the recruitment cycle, which focuses on acquiring human intelligence (HUMINT). Prospective targets are considered especially noteworthy based on two key factors: their access to valuable information and their personal motivations to commit treason against their own government or organization.

This recruitment cycle is universal across the world's various intelligence services and consists of the following elements:

- Spotting – identifying potential targets of recruitment
- Assessing – determining a target's access to valuable information
- Developing – building rapport with the target to understand his/her motivations

- Recruiting – pitching the target to clandestinely provide the information and become an asset (source) for the case officer
- Handling – working with the asset to provide direction and to securely pass the information
- Terminating – securely ending the working relationship between case officer and asset

CARVER's role can be useful in each phase of the recruitment cycle. For example, when a case officer is looking to identify potential targets, he/she can run through the CARVER criteria to determine suitability. Following is a list of questions, utilizing the CARVER acronym, which the case officer could use to qualify a prospective intelligent asset (They are by no means the only questions; each situation is different):

Criticality

- Is this person critical to my statement of requirements?
- Is this person critical to the overall organization that I am looking to acquire information about?
- Is this person critical to someone else that I am targeting within the organization?

Accessibility

- How much direct access does this person have, to the information I am looking to acquire?
- Does this person have access to someone or something else that can assist me?
- How accessible is this person to me (the case officer)?

Recoverability

- If our working relationship became exposed, would the asset be able to recover personally and professionally?
- If our working relationship became exposed, how difficult would it be for the relationship between the organizations involved to recover (example: United States and a foreign government)?
- If we lost access to this asset, or the asset lost his/her access to valuable information) how long would it take for us to recover by finding new access?

Vulnerability

- Does this person have vulnerabilities or weak points that can be exploited?
- Is there a vulnerability within the organization I am looking to obtain information about?
- Am I (the case officer) vulnerable to exposure by meeting with this person?

Effect

- What would be the positive effect of recruiting this individual?
- What would be the negative effect if our relationship became compromised or exposed?

Recognizability

- How recognizable is this person to me as someone with access?
- If this person became a source, would it be easily recognizable by others?
- How recognizable is this person by others and would that affect public encounters?
- How effective would a specific strategy be for attempting to recruit this person?

It should be noted that another popular technique for determining a person's motivation to commit espionage is to consider their personality against the acronym MICE, which stands for Money, Ideology, Coercion and Ego. It should come as no surprise that money is the most commonly used approach to get people to spy, while ideology is the most effective. People who spy out of ego can be dangerous (aka loose cannons), while people who are coerced into spying (i.e., blackmailed) are the least effective.

Note: The CIA does not use coercion as a recruitment method, primarily out of ethics but also because of its unreliability as a technique.

CARVER in Strategic Targeting Operations

Now that we have addressed the issue of individual recruitment (which can be considered a tactical piece of an intelligence operation), it is important to look at the use of CARVER in more strategic

CARVER and NWWAM as Sales Tools

Just as a security professional would complete a CARVER matrix worksheet when assessing the vulnerabilities of a critical infrastructure, a case officer would perform a similar assessment on a potential human target. Contrary to popular belief, CARVER is not some magical formula that is proprietary only to the intelligence community. In fact, business schools around the world – not to mention thousands of used car salesmen – have been utilizing a similar method to qualify prospective clients for their commercial sales products. The selling acronym of NWWAM has been around for decades and is a technique that has been equally effective in identifying human weaknesses. It stands for:

Needs – Does the prospect need this product?
Wants – Does the prospect actually want the product?
Wounds – Is there something in this prospect's past that I, as the salesman, can leverage to my selling advantage?
Authority – Is the person the correct decision-maker with the authority to sign off on the purchase?
Money – Does the person have money to pay for the product I'm selling?

Whether you're selling someone on the idea of buying a toaster or engaging in espionage, the principles of human behavior are no different. In sales or espionage, everyone has a unique set of beliefs that can be exploited.

intelligence terms. More specifically, let's look at how CARVER can be used for larger strategic targeting operations. The intel community defines the word targeting as, "a proactive process designed to generate offensive operational action against a target." Translated, it means targeting is a combination of analysis and operations that collectively results in taking action against an enemy.

When intel-community people talk about conducting a targeted campaign, they might be referring to disrupting or dismantling a terrorist organization such as al-Qaeda or ISIS. This campaign, or mission, might include functions such as organizing, collecting and analyzing information with an objective to produce the intelligence needed to identify, capture and/or destroy their targets. To most effectively carry out their responsibilities, the targeters (analysts with an operational focus) look to bridge offensive operations by identifying the adversary's vulnerabilities, choke points, information gaps, recruitment leads and key personnel. Time and immediate threats from the enemy might dictate the options available to the targeter. Operational targeters are different from analysts in that their focus is on delivering supporting intelligence to on-the-ground operators – ideally in real time – rather than high-level policymakers.

For an operational targeter, knowing the mission is critical. Unlike traditional analysts sorting through available intelligence with the aim of drawing a big-picture hypothesis, the targeter's objectives are much more specific. Therefore, a targeter must be prepared to support an action by offering more effective questions to collectors, give exploitable data on target vulnerabilities and provide leads for operations against threats.

Typically, six steps constitute the targeting process, all of which can utilize CARVER to assist with operational analyses:

Step One: Select a target. Target selection is often based on the needs of the policymakers. Essentially, the targeters must be very clear about the specifics of the target, because vague descriptions or loose requirements can lead to incomplete or inconsequential intelligence.

Step Two: Define mission objectives. When the target has been clearly identified, it is important to outline the parameters of the operation. This might include defining the threat, penetrating the target's organization and determining how action might be taken against the target.

Step Three: Determine what is already known. Incorporating the same logic as the CARVER methodology, an initial target analysis is conducted against the target (and remains continuous throughout the operation). During this step, the targeter will assemble, organize and analyze all data to identify gaps for additional collection.

Step Four: Take inventory of existing collection/action mechanisms. Just as CARVER is useful for characterizing and ranking assets, here the targeter will take inventory of his/her available intelligence capabilities and identify the means to leverage the resources to advance the mission. Such inventory characterization could include existing relationships with other government agencies, access to foreign liaison assistance, available clandestine human assets, technical capabilities and action authorities granted.

Step Five: Plan new operations against the target based on

analysis. In this step, the targeter develops a concept of operations (CONOPS) for advancing and achieving the mission objectives. The plan might include ideas for performing non-alerting collection strategies, developing new assets, tasking existing assets, engaging in new investigations and conducting covert action to disrupt or negate the target.

Step Six: Obtain approval and begin execution of plan. After the targeter has gathered all the necessary information, built target packages for assets and adversaries, and developed a CONOPS, the plan must still be approved for use. The targeter should know his/her approval audience; what, if any, coordination issues might arise with external partners, and consider the common-sense limitations on action operations.

Just as a security professional might deliver a final vulnerability-assessment report for a critical infrastructure, a targeter will create a package on an individual, group or physical target. When this component is complete, a CONOPS will be formulated and submitted for approval. For example, by using CARVER criteria to justify his/her findings, the targeter might analyze that:

> *The probability of attack against a railway station is significantly higher than an airport and that resources should be dedicated to investigating who might have direct access to rail stations and/or might have been seen conducting possible casing or surveillance.*

Likewise, the targeter could use CARVER for HUMINT recommendations. An example:

> *Based on my assessment of the known supporters and sympathizers to this terrorist organization, I believe that Mohammed is the most susceptible person to potential recruitment and I would suggest appealing to his moral and ethical dilemma about terrorism as a possible motivation to exploit.*

By using strategic analysis principles found within the CARVER methodology, a targeter can help explain what is happening now and what is likely to happen in the future, based on the information at hand.

8. Utilizing CARVER for Red Team / PENTEST Operations

Never test the depth of the river with both of your feet.
—Warren Buffett

IN THE PARLANCE OF THE MILITARY, a red team is an independent group that challenges an organization to improve its effectiveness. Both the CIA and the military special-operations community have used red teams for decades to uncover unknown physical vulnerabilities, as well as to test the response capabilities of force protection units, first responders or other internal/external security measures. The process has also been used to predict the future actions of others, whether they be foreign government leaders, terrorist groups/criminals, business competitors, political opponents, or some other adversary. By attempting a series of attacks, the red team "aggressors" challenge current standard operating procedures and test the ability of the system to mitigate and respond to threats, all within a safe and controlled environment.

When performing a target analysis and vulnerability assessment on a critical infrastructure, the red teaming process may be incorporated by the assessment team as a method to expose weaknesses within the infrastructure's security system. At SMI, we have frequently used red teams to assist with identifying potential areas of weakness. Red teams also help demonstrate to the client the severity of some of these weaknesses. For example, some clients might hire a consultant to conduct a vulnerability assessment because it is part of a mandated federal regulation. As such, they might not actually incorporate any of the security recommendations offered by the consultant due to budget constraints. However, if the consultant can physically demonstrate that he/she penetrated the client's existing security measures, the client will be motivated to address the issue. This is another reason that red-teaming is known as penetration tests – or PENTESTs – in the business world.

At SMI, we have conducted various PENTESTs over the years. In one instance, we were asked to "break in" to a client's facility (which was a large, privately owned utility) to see if we could defeat their existing security measures. Instead of circumventing security to conduct a physical attack against the venue, however, the client requested we attempt a covert intrusion and engage in economic espionage collection. We were required to try and plant hidden listening devices within their facility and to see if we could access their sensitive files or conversations. For a team of former intelligence operators, we welcomed the challenge.

Our attack planning began with surveillance against the facility, which included taking numerous photos and videos of the property, following employees home from work, making phone calls to employees at their desks, under the guise of the IT department, in an attempt to elicit sensitive information, and even inserting one of our team members into the building as a phony telephone repairman. We conducted exhaustive open-source due-diligence to determine which employees might be susceptible to recruitment through money, ideology, coercion or a bruised ego/revenge (often referred to in the intelligence world as that aforementioned acronym MICE). The planning phase, in which we measured our options against the variables of the CARVER Methodology, lasted about one week.

The following chart depicts how we used CARVER as part of our surveillance phase:

OBJECTIVE: Covertly Plant Audio Devices In Secured Spaces

ASSET	DESCRIPTION	Criticality	Accessibility	Recoverability	Vulnerability	Effect	Recognizability	RISK	Pa
	Public Areas: Lobbies,Cafeteria, Elevators	2	4	2	5	3	5	21	0.70
	Executive Conference Room	4	3	4	2	4	3	20	0.67
	CEO's Office	5	1	5	2	5	1	19	0.63

CARVER Color Coding

24+	SEVERE
15-23	MODERATE
6-24	LOW

CARVER Ranking

1 – Much Lower Than
2 – Lower Than
3 – The Same As
4 – Greater Than
5 – Much Greater Than

As you can see from the matrix above, the most likely probability of attack (Pa) for covertly planting audio devices within the target's office space would be in the public areas. The most difficult location would be in the CEO's office. During the actual operation, our operatives circumvented security measures and planted surveillance devices at all three locations.

In the end, the CARVER matrices determined we could best achieve our mission via a series of training courses open to the public. The client had been offering distinct engineering courses for other professionals within the industry. A former international intelligence officer and I signed up as students under aliases; that was the easy part. Next, we assigned some of our surveillance team members to follow a certain employee into Starbuck's, where he stopped every morning for coffee prior to driving through the plant gates. We chose this particular individual because he always wore his company ID badge/door access card on a lanyard, which hung conspicuously from his belt. One team member stood behind him in line and "accidently" bumped into him, brushing a concealed RFID card reader against the man's access badge. Without his knowledge, our team was able to clone his access card.

When the day of the course arrived, my colleague and I passed through security with our false identification and were led into the training room. Because we were 30 minutes early, we were allowed to venture to the nearby cafeteria for coffee with our red "Visitor" badges prominently displayed around our necks. I made a detour into the men's room, where I changed into work clothes that matched the typical dress uniform of the facility's employees. I also dawned a fake ID badge identical to the company standard. Using my duplicated door card, I accessed the stairwell with a swipe and ventured up to the next three floors. As I quickly moved through the hallways during the early morning hours, I discretely stuck tiny listening devices under the desks in random empty cubicles.

After ten minutes of planting my bugs, I changed clothes and returned to the classroom. My colleague was nowhere to be seen. She had also engaged in the operation, albeit via different means. As an attractive young woman, she used her feminine powers of persuasion to gain access to the executive elevator. After buying two large cups of coffee in the cafeteria, she stood outside the elevator with her hands full. When she noticed a male employee about to board, she expertly played the damsel in distress and asked him to hit the top-floor button. The chivalrous man swiped his access card and granted her request. When she got off the elevator, alone on the top floor, she ditched the coffee and made her way through the C-Suite with a confident demeanor that did not raise eyebrows among the employees checking morning emails. Like a true professional, she planted her audio devices under the executive conference table as well as the bookcase in the CEO's office.

Within a matter of minutes, she was back sitting on the opposite side of the classroom from me (we pretended not to know each other) preparing for class. We exchanged text messages to inform each other of our accomplishments. Because we now had each floor wired for sound, we decided to sneak away from class after the first break.

After an hour nervously waiting to see if the police would barge through the door, the instructor let everyone out for a coffee break. Before my colleague and I headed for the parking lot and our rental cars, we had one more task to complete. We returned to the cafeteria and plugged what looked like an everyday air freshener into one of the wall sockets. Yet, this was no ordinary air freshener. It was actually a Wi-Fi sniffer, which could clandestinely collect cellphone and laptop signals. Within another 20 minutes, however, we were back at our hotel, relieved to have not been caught.

The next morning, we debriefed our clients: the CEO, COO, security director, HR director and an in-house counsel. Needless to say, the panel was shocked and unhappy with the ease at which we had infiltrated their organization. While the CEO bemoaned the results, my colleagues and I recognized that CARVER was the foundation on which we had built our ops plan.

9. Utilizing CARVER for Strategic Business Management Analysis

The essence of strategy is choosing what not to do.
—Michael Porter

Up to this point, we have demonstrated how CARVER can be used for security assessments and operational-planning purposes. Therefore, it should come as no surprise that as a multi-dimensional tool, CARVER can also be used in the business world. Strategic management has always been – and most likely always will be – buzz words in the business community. And while the adage, "business is war" has become a trite and absurd expression to describe the competitive business environment, there is no denying that companies are constantly measuring their internal processes, as well as studying those of their competitors. In this chapter, we will focus on how private-sector organizations measure their operational procedures, decipher the marketplace, assess external threats and make better business decisions, all with the help of the CARVER Methodology.

Sadly, management consultants have often been lumped in the same category as lawyers. They have been the butt of many jokes, especially because of their obscene hourly billing practices, which often leave their clients scratching their heads and asking themselves, "what did I just pay all that money for?" Firms such as Bain, McKinsey, and Boston Consulting Group have been the dominant leaders in the strategic-management field for decades. Harvard MBA-types, dressed in well-appointed suits, who dazzle their clients with fancy business jargon and colorful PowerPoint slide decks, are typical in the industry. This is not to say that management consultants are charlatans, eagerly preying on their "innocent" Fortune 500 clientele. On the contrary. Management consultants bring something to the business world that perfectly translates to the security industry, as well as many other disciplines. That is their ability to document, measure, and assess their clients' vulnerabilities and competitive strengths, while gauging the current business environment and anticipating future trends. This is exactly what the CARVER methodology does.

Utilizing CARVER in the global business arena, companies can determine some of the following:

- Which business lines are critical?
- What market share is accessible?
- How quickly can they financially recover from bad PR?
- Where are their competitors vulnerable?
- What is the effect of running a particular ad campaign?
- What clients are recognizable?

You get the idea.

Just as the infamous SWOT analysis taught millions of MBA students that they should view the core of a business through the criteria of Strengths, Weaknesses, Opportunities, and Threats, the CARVER methodology echoes that sentiment while extending the lens from which a business can be analyzed. For example, if we compared the differences between a SWOT analysis and a CARVER analysis, using the same business case study, we would find that we can obtain more detailed information about the company utilizing CARVER than SWOT. Let's breakdown the following fictitious company using both models.

The ABC Widget Co.

The ABC Company of Atlanta has been the market leader in selling "widgets," inventing the product just four years ago. They sell these widgets at $25 each. After all their overhead costs are figured in, the ABC Company makes a net profit of $7 off every sale. They sell approximately 100,000 widgets per year, leaving them with an annual net profit – before taxes – of $700,000. Within the past 12 months, two new competitors have started offering a similar yet less-expensive version of the widget. One of those competitors has a deep-pocketed investor who is funding an aggressive advertising campaign. Their widget sells for $19. The other company manufactures an inferior "budget" widget in China which, as a result, sells their product for only $15.

SWOT:

Strengths: Brand awareness Marketing leader Industry experience Strong profit margins Inventor of product	**Weaknesses:** Open to competition More expensive to build More expensive for consumer
Opportunities: Higher quality product "Made in America" product Can establish image as "Rolls Royce" of widgets	**Threats:** Cheaper competitors steal market share Deep-pocketed investor provides marketing dollars/R&D

As you can see, CARVER is more detailed in breaking down strengths, weaknesses, opportunities and threats than the traditional SWOT analysis model. In developing the following CARVER matrix, it was found that estimating the probability of attack was not necessary – though calculating the risk to the product was needed. The threat of cheaper competitors and deep-pocketed investors provided the biggest risk to ABC Widgets.

OBJECTIVE: ABC Widgets Competitive edge

ASSET	DESCRIPTION	Criticality	Accessibility	Recoverability	Vulnerability	Effect	Recognizability	RISK
Brand awareness, Market leader, Industry experience, Strong profit margin, Inventor of product	Strengths	4	3	3	4	3	4	21
Open to competition, More expensive to build, More expensive for consumer	Weaknesses	5	3	2	4	4	4	22
Higher quality product, "Made in America" product, Can establish image as "Rolls Royce" of widgets	Opportunities	4	2	4	2	5	1	20
Cheaper competitors steal market share, Deep-pocketed investor provides advertising dollars / R&D	Threats	5	4	4	4	4	4	25

CARVER vs. GE's Five Slides

So, if you think we are pushing this CARVER thing a little too far, here is a real-life example that might help demonstrate the methodology's value in the business world. Jack Welch, the somewhat controversial former CEO of General Electric (whom many consider the greatest CEO of the past 50 years), used a similar method to CARVER to assess his company. Welch called it "The Five Slides."

The Five Slides were essentially five strategic questions – with several sub-questions included under each slide – that GE executives were required to ask themselves each quarter. Much like a CARVER assessment, the criteria of the questions examined the current situation, along with anticipated outcomes. Below are GE's "Five Slides" (strategic questions) along with their CARVER equivalent.

Slide One: What does the playing field look like now?

Who are the competitors in our business, large and small, new and old?

Who has what share, globally and in each market?

Where do we fit in?

What are the characteristics of this business?

Is it commodity or high value or somewhere in between?

Is it long cycle or short?

Where is it on the growth curve?

Where are the drivers of profitability?

What are the strengths and weaknesses of each competitor?

How good are their products? How much does each spend on R&D?

How big is each sales force?

How performance-driven is each culture?

Who are this business's main customers and how do they buy?

These questions could fit under the criterium of Criticality in the CARVER matrix.

Slide Two: What the Competition Has Been Up To?

What has each competitor done in the past year to change the playing field?

Has anyone introduced game-changing new products, new technologies, or a new distribution channel?

Are there any new entrants, and what have they been up to in the past year?

These questions could fit under the criteria of Accessibility and Effect in the CARVER matrix.

Slide Three: What Have We Been Up To?

What have you done in the past year to change the competitive playing field?

Have you bought a company, introduced a new product, stolen a competitor's key sales person, or licensed a new technology from a start-up?

Have you lost any competitive advantages that you once had – a great salesperson, a special product, a proprietary technology?

These questions could fit under the criterium of Recognizability in the CARVER matrix.

Slide Four: What's Around the Corner?

What scares us the most in the year ahead – what one or two things could a competitor do to nail us?

What new products or technologies could your competitors launch that might change the game?

What M&A deals would knock us off our feet?

These questions could fit under the criteria of both Vulnerability and Effect in the CARVER matrix.

Slide Five: What's Our Winning Move?

What can we do to change the playing field – is it an acquisition, a new product, globalization?

What can we do to make customers stick to us more than ever before and more than to anyone else?

These questions could fit under the criteria of both Effect and Recoverability in the CARVER matrix.

How the Great Management Gurus View Statistical Methodologies

Peter Drucker, the godfather of strategic management, was famously quoted as saying, "If you can't measure it, you can't manage it." Similarly, Pearson's Law states that, "When performance is measured, performances improves. When performance is measured and reported back, the rate of improvement accelerates exponentially." This focus on benchmarking and measured progress is at the heart of improvement. Thus, the need to assess.

What professionals need to recognize is that performing a CARVER assessment is not a one-and-done solution. Assessing infrastructure, business processes, or even people must be an ongoing event. The initial CARVER assessment will provide you a baseline, which should be re-evaluated on a regular basis to determine improvement and regularly expose weaknesses. Strategic management, much like security management, is organic. Meaning that it should be utilized daily. It should not be an idea, a manual, or a process that is left to collect dust on a shelf. One of the biggest fallacies of successful management is that having a solid plan or strategic vision is more important than implementing those tactics on a consistent basis. In essence, heavy strategic planning can actually kill useful strategic implementation.

Even Edwards Demming, the man who revitalized the Japanese auto industry following World War II, cautioned engineers that a failure to properly identify the underlying cause of a problem could greatly skew statistical information gathering procedures and lead to inaccurate assessments. In the book *The Essential Demming: Leadership Principles of the Father of Quality*, Demming discussed the causes of inaccurate statistical gathering methods. He stated:

> *One of the important uses of statistical techniques is to help an engineer or scientist to distinguish between two types of cause, and hence to fix (with adjustable risk of being wrong) the responsibility for correction of undesired variability or of undesired level.*

> *Confusion between common causes and special causes is one of the most serious mistakes of administration in the industry, and in public administration as well. Unaided by statistical techniques, man's natural reaction to trouble of any kind, such as an accident, high rejection-rate, stoppage of production, is to blame a specific operator or machine. Anything bad that happens, it might seem, is somebody's fault, and it wouldn't have happened if he had done his job right.*
>
> *Actually, however, the cause of trouble may be common to all machines, or the fault of management. Demoralization, frustration, and economic loss are inevitable results of attributing trouble to some specific operator, foreman, machine, or other local condition, when the trouble is actually a common cause, affecting all operators and machines, and correctible only at a higher level of management.*

What Demming is trying to say is that although the assessor might identify fault in the operator, the actual cause of the problem might not be ineptitude of the operator (i.e., human error) but instead inadequate system polices and/or poor employee training procedures. For example, if during a security survey a security guard is deemed as susceptible to social engineering or elicitation techniques, the vulnerability lies not with the guard himself but with the policies for training him to recognize these forms of manipulation. Therefore, identifying this vulnerability during a CARVER assessment would mean highlighting the deficiency in training and management procedures, rather than the vulnerabilities of the guards themselves.

Whether in business or in life, progress should be measured frequently. Performing assessments are an objective way to determine where you are going and where you have been, while keeping you apprised of what your current challenges and vulnerabilities might be. By understanding where you need to improve, you can best manage successful change.

10. Utilizing CARVER with a Cyber Security Assessment

If you spend more on coffee than on IT security, you will be hacked. What's more, you deserve to be hacked.

—Richard Clarke

ACCORDING THE Department of Homeland Security's Cyber Security Division, "Cyber adversaries have presented a full spectrum of threats not only to the U.S. government, but also to private organizations and critical-infrastructure sectors." For this reason alone, we recommend that in most instances critical infrastructure/physical-vulnerability assessments be coupled with a comprehensive cyber-security assessment. Acts of physical destruction (i.e., aggressive attacks by outsiders or life-threatening accidents/natural disasters) could bring unwanted public attention to an organization, but the thought of a crippling cyber attack also keeps public- and private-sector leadership awake at night.

In much the same way as a traditional CARVER target-analysis engagement begins, a cyber-security assessment aims to answer the following key questions:

- What assets are we trying to protect?
- What are the risks to these assets?
- How well does the security solution mitigate those risks?
- What other risks does the security solution cause?
- What costs and trade-offs does the security solution impose or create?

Because the cyber universe is so vast, determining the initial objectives of a cyber-security assessment can appear daunting to those of us who are not full-blown techies. There is no question that the landscape of electronic vulnerabilities continually changes at lightning speed. New threats emerge daily. Therefore, if security professionals are not deeply entrenched within the nuances of the cyber-security field, they might seem constantly at a disadvantage to their adversaries. Although this is true to some extent, it does not mean all hope is lost for the security professional who is familiar only with the physical aspects of security.

In many ways, conducting a cyber assessment requires the same strategic mindset as a physical assessment. You are still looking at the system through the eyes of your adversary. A terrorist might follow the steps outlined in the U.S. intelligence community's terrorist planning cycle (including identifying a target, performing surveillance, acquiring materials, etc.), but a hacker would also follow a certain set of steps for attacking/compromising IT assets. Such a process might involve:

- Acquiring targets
- Gathering information and data
- Reconnoitering a network
- Discovering vulnerability
- Compromising the target
- Controlling the host
- Expanding control
- Covering tracks

Although the infrastructure might be different, the concept is similar.

At SMI, we have made it a point to conduct joint assessments that examine both the physical and cyber vulnerabilities of a system, to ensure that we are providing the most comprehensive assessment report possible to our clients. We proudly tout our associates as some of the best in the business when it comes to understanding explosive blast effects, counterterrorism and other acts of physical sabotage, but we are not too proud to call on the expertise of others when necessary. On more than one occasion, we have relied on trusted strategic partners to ensure we are not overlooking important cyber vulnerabilities. To complement the talent of our partners, and maintain a consistent report format for our clients, we utilize the CARVER methodology to demonstrate threats and vulnerabilities. The principles of deter, detect, delay, respond and mitigate still ring true in an online world, and a CARVER scoring matrix is easily transferable from physical assets to virtual assets.

Just as a bomb technician examines the vulnerabilities of a building from the threat of an explosive device, a cyber-security expert reviews the components that make up the IT infrastructure and keeps an eye on how a hacker might compromise it. Just some of the areas that could be considered include:

- Open-source intelligence
- Network connectivity
- Dedicated security devices
- Remote access services
- Network operating systems
- Disaster recovery and planning
- Policies and procedures

This list might sound exhaustive or even intimidating, but to those unfamiliar with the cyber-security world, often the biggest IT network vulnerabilities are the people who use them. Employees who employ overly simplistic passwords, click on non-work-related email attachments and visit unauthorized websites, typically constitute the greatest threats within any organization. For this reason alone, it is important to recognize that regardless of the network's security, the human factor is still the time-tested gateway to accessing the system.

A perfect example of how we combine a physical assessment with a cyber assessment is demonstrated as part of our work with the North America Energy Reliability Council (NERC) Critical Infrastructure Protection (CIP) requirements. As mentioned previously, SMI has extensive experience conducting the physical assessments under regulation CIP-014. To complement the other CIP requirements, however – specifically CIP-10, which mandates that all power utilities "perform a vulnerability assessment on cyber assets and provide documentation" – we don our IT hats and approach the problem from the standpoint of a cybercriminal.

To ensure that our clients are compliant with NERC guidelines, we address each requirement individually. For example, CIP-007 calls for "a cyber vulnerability assessment of all cyber assets within the electronic security perimeter (ESP), including a document identifying the vulnerability-assessment process, a review to verify that only the ports and services required for operation of the cyber assets within the ESP are enabled, a review of controls for default accounts, and documentation of the results, mitigation plan and mitigation status." It might sound like geek-speak, but essentially NERC offers its users a compliance roadmap to follow.

Furthermore, NERC CIP-007 provides these requirements:

R8. Cyber Vulnerability Assessment. The Responsible Entity shall perform a cyber vulnerability assessment of all Cyber Assets within the Electronic Security Perimeter at least annually. The vulnerability assessment shall include, at a minimum, the following:

- R8.1. A document identifying the vulnerability-assessment process
- R8.2. A review to verify that only ports and services required for operation of the Cyber Assets within the Electronic Security Perimeter are enabled
- R8.3. A review of controls for default accounts
- R8.4. Documentation of the results of the assessment, the action plan to remediate or mitigate vulnerabilities identified in the assessment, and the execution status of that action plan

This balance between physical and cyber allows the vulnerability-assessment team to develop a more complete picture of the potential weaknesses within the energy system. It should also be pointed out that a third factor (not included as part of NERC regulations but certainly no less important as a vulnerability) is the counterintelligence/counter-espionage/insider threat.

It has been our experience that the most exploitable gap in any organization is the human one. Unfortunately, due to legal and human-resources considerations, it is more difficult to perform an assessment on individuals within the organization who might have the most to gain by disrupting operations from within (please refer to the previous chapter on PENTESTs for further information).

Defining Cyber Threats

It is undeniable that U.S. infrastructure is under increased threat from malicious cyber actors, whether they be hostile foreign nations/intelligence services, international/domestic terrorist organizations, or private hackers seeking to acquire recognition and/or financial compensation. Exploiting an organization's system or network could result in the loss of sensitive data (i.e., intellectual property, personal information, research and development material), and denial/disruption of business operations. It could also have severe reputational impacts. A successful attack could literally cost an organization billions of dollars. Unfortunately, many of the tools available to instigate such an attack are widely available on the Internet, while external access points (e.g. modems, firewalls, wireless connections, portable devices, media) are readily exploited. Viruses, worms and other Trojan Horses could allow individual employees to unknowingly invite malicious individuals into their network.

For industrial organizations and/or utilities, such as water, power, railways and even nuclear facilities, Supervisory Control and Data Acquisition (SCADA) systems might still be in operation. These antiquated – more mechanical – platforms are often decades old and rely on a combination of interdependencies between dated and new technologies to operate the system architecture. Integrating Internet-based protocols into a wide-area communication system (a system that was developed before the World Wide Web ever existed) has provided an increased challenge to security professionals looking to defend against cyber-attacks.

It is important to point out that some security professionals argue that because SCADA systems were built on an independent network many years ago, they are easier to recover should they be attacked.

Utilizing CARVER to Organize Your Cyber Assessment Finding

One of the key differences between a cyber-vulnerability assessment and a physical-vulnerability assessment is the number of tools that can be deployed to test the system. Often, in a physical assessment, the team members must depend on their own in-person inspection of assets to test the reliability of doors, windows, alarms, access-control panels, CCTV cameras, etc. With a cyber assessment, however, the team can engage a series of scanners and other tech tools to exploit the network at a distance. This does not mean that a cyber assessment is somehow less subjective – or less personal – and does not require a methodology for quantifying results. In fact, the opposite might be true. Because a cyber assessment encompasses so many various assets (e.g., ports, firewalls, CPUs, software, etc.), a greater emphasis might be required to rack and stack so many items of concern. Once again, CARVER lends itself as a useful instrument to organize these findings.

By engaging the CARVER methodology, the assessment team can better maintain a meticulous inventory of assets while establishing a benchmark from which to offer improvements.

The purpose of this book – much less this chapter – is not to provide an all-encompassing cyber blueprint (there are volumes of books out there that cover thousands of IT security steps), but we do want to reiterate the value of CARVER as a means for prioritizing what could be an overwhelming amount of IT assets as part of a cyber-vulnerability assessment.

Many cyber-assessment tools define risk and/or probably of attack using a scoring matrix of Low-Medium-High (sometimes: Least-Minimal-Moderate-Significant-Most), but CARVER allows both the assessor and the decision-maker to have a more concrete scoring criteria. This specificity allows for tighter controls to be enacted within future policies and procedures, as well as moving away from a basic, one size fits all model.

On the next page, we have included a sample of how CARVER can be used for a cyber-vulnerability assessment.

As you can see, cyber-security vulnerability assessments can be equally as important as physical-security vulnerability assessments. Fortunately, the CARVER methodology lends itself to a consistent, logical, format, which can be standardized across industries, systems, sub-systems and assets. design basis threats (DBTs) must still be considered to assist the assessment team in determining the probability of attack (Pa).

Just as in the world of physical security, the overlying objective of IT security is to prevent unauthorized intrusion and attack. As such, our philosophy of deterring, detecting, delaying, responding, and mitigating holds true. The integrity of, and access to, a cyber arena is no different than a physical arena. The underlying difference, perhaps, is providing confidentiality to the IT user. Regardless of definition, there is no denying the need for cyber assessments or the need for a comprehensive methodology such as CARVER to prioritize the threats.

ASSET	DESCRIPTION	DESIGN BASIS THREATS	Criticality	Accessibility	Recoverability	Vulnerability	Effect	Recognizability	RISK	Pa
Servers	Data Management	Trojans, Rootkits, Virus-Port Scanners, Network Mappers	5	3	4	4	5	4	25	0.83
Mainframes	Data Management	Trojans, Rootkits, Virus-Port Scanners, Network Mappers	5	3	4	4	5	4	25	0.83
Mobile Devices (Phones)	Personal/ Company Provided	Apps, Email Use – Loss of Device and Sensitive Information	4	4	4	5	4	4	25	0.83
Password Management	Provides Secure Access	Loss, Social Engineering, Phishing, Remote Password Crackers	4	3	3	4	4	2	20	0.67
Desktops, Laptops, Tablets	Personal/ Company Provided	Apps, Email Use – Loss of Device and Sensitive information	4	3	4	4	3	4	25	0.73
Email Services	Company or Public Service	Improper Use of Each	3	4	3	4	3	3	20	0.67
Software Updates and Patches	Security Patches	Hackers Look for Non-Compliance	5	4	3	3	4	4	23	0.77
Removable Media	Thumb Drives, CDs	Lost or Stolen	4	4	3	3	3	3	20	0.67
WiFi Routers	Weak SSIDs	System Access	4	3	3	4	4	4	25	0.73

11. Utilizing CARVER for Operational Planning

The enemy invariably attacks on two occasions: when they're ready and when you're not.

—Murphy's Military Law #15

Planning

Up to this point, we have demonstrated the value of CARVER as a target analysis and vulnerability-assessment methodology. But CARVER is also a useful tool for operational planning. Thanks to its initial development as an offensive targeting process by the U.S. Army Air Forces during World War II, CARVER can still be used effectively today by numerous military and intelligence professionals for planning sensitive operations. CARVER was even rumored to have been involved in certain planning aspects of the assault of Osama bin Laden's compound in Abbottabad, Pakistan, by U.S. Navy SEALs in 2011.

To understand the finer aspects of operational planning, and how CARVER can be incorporated into an organization's needs, it is necessary to define what planning is. Furthermore, it is necessary to describe how – and where – CARVER fits in to the operational planning process.

Plans vs. Planning

In its simplest terms, planning is a scheme, program or method worked out beforehand to accomplish an objective. But planning alone does not guarantee flawless execution. Just because you have a plan to secure critical infrastructure from terrorist or other nefarious attack, it doesn't always mean the plan will be successful. You need to take additional factors into consideration, such as exercising/rehearsing the plan, training the team members on their roles and responsibilities, passing accurate internal and external communications during an event, coordinating with liaison partners, compartmentalizing sensitive information, and ensuring that contingencies are activated, just to name a few. Even then, you will still only be able to anticipate – rather than know – what is going to transpire next. In fact, plans and planning are two entirely different things.

As General Dwight D. Eisenhower so famously stated, "In preparing for battle, I have always found that plans are useless but planning is indispensable." Translation: A detailed game plan rarely unfolds with flawless execution – particularly when you are facing a difficult adversary, whether known or unknown. You should always emphasize understanding every possible contingency ahead of time by asking the simple question, "What if?" This is the type of planning necessary to ensure that a critical infrastructure is properly secured from adversarial attack.

With all due respect to General Eisenhower, I have always preferred the quote from former heavyweight boxing champion Mike Tyson, who said, "Everyone has a plan to beat me, until they get punched in the nose." Those with the ability to understand thoroughly the intricacies of the planning process are the ones who can adapt and overcome when faced with a difficult response situation. I sum up this ability to adapt in two words made famous by the Boy Scouts of America: Be prepared.

Preparation

Preparation is defined as the state of readiness. Anyone who has ever been tasked with securing a facility, particularly in a war zone or other high-threat environment, understands this concept. As such, the individual(s) responsible for security must operate in a state of constant alertness, keenly aware that Mr. Murphy, of Murphy's Law fame, can rear his ugly head at any moment.

In traveling to over 130 countries and working with security professionals from various military, law enforcement and intelligence services, I found nearly all of them familiar with Murphy's Law. When asked to define Murphy's Law, most people will respond with the phrase, "Anything that can go wrong, will go wrong." This is true, but most people never learn the origin, or additional adages, behind Murphy's Law.

The Origins of 'Murphy's Law'

According to the Edward's Air Force Base newspaper Desert Wings, "Murphy's Law was born at Edwards Air Force Base in 1949 and named after Captain Edward A. Murphy, an engineer working on Air Force Project MX981, (a project) designed to see how much sudden deceleration a person can stand in a crash." The article went on to claim that, "One day, after finding that a transducer was wired wrong, Murphy cursed the technician responsible and said, 'If there is any way to do it wrong, he'll find it.'" The contract's project manager kept a list of Captain Murphy's laws and claimed this as number one. Interesting enough, the Russian military has a similar law, known as the Law of Buttered Bread. In this version, the Russians claim that a piece of bread buttered on one side will always land buttered side down on the floor.

Here are Captain Murphy's top five laws, which are applicable to the security-planning process and will help you to expect the unexpected:

1. In any field of endeavor, anything that can go wrong will go wrong.
2. Left to themselves, things always go from bad to worse.
3. If there is a possibility of several things going wrong, the one that will go wrong is the one that will cause the most damage.
4. Nature always sides with the hidden flaw.
5. If everything seems to be going well, you have obviously overlooked something.

Aside from being an interesting backstory, Murphy's Laws are essential to the planning process. Every CARVER assessment, utilized whether for offensive or defense purposes, must take in consideration Captain Murphy's thought process. This means the plan writer and/or the planning committee must be objective, realistic and humble about their goals and capabilities. More important, they should always be ready to answer the question, "What if?"

The Five Assumptions of a Crisis

At our firm, Security Management International, we recognize that Mr. Murphy is ever-present and has plenty of relatives overseas. In fact, it has been our experience that the further one moves away from the flag pole (i.e., an organization's headquarters or base of operations), the more likely the Murphy clan will materialize. Therefore, at SMI, we always stress the 5 Assumptions in Any Crisis:

Assumption 1 – The problem is always worse than it appears. This means the crisis will usually grow before it shrinks, and our people should avoid blowing off initial intelligence as "So what?"

Assumption 2 – There are very few true secrets in the world. Assume everyone will eventually find out. In other words, be cautious of denying something that could come back to bite you later. Also, think before you speak in a crisis. Your emotions could lead you to say something you'll regret later.

Assumption 3 – You and your handling of a crisis will be portrayed in the worse possible light. The media are not your friends during a crisis. Accept this, and defuse them quickly. Your public message is even more critical in the early stage of a crisis. Understand that a lack of visibility will be seen as an admission of guilt.

Assumption 4 – There will be changes in processes and people at the end of the crisis. This means no crisis ends without blood on the floor. Someone will be blamed and made an example of (except in the government, where the person at fault will probably be promoted).

Assumption 5 – The organization will survive, ultimately stronger for what happened. Although it is not a pleasant situation, there is not a crisis you cannot learn from. If you have the right leadership, your organization should ultimately grow stronger in the long run.

The Planning Process

As the father of modern-day automotive processes and measurement standards, W. Edwards Deming (the American management consultant who helped build Japan's auto industry after World War II – and is still celebrated in Japan with the Deming Award), famously stated, "If you can't describe what you are doing as a process, you don't know what you're doing."

Operational planning is a process. No military general – or security director for that matter – worth their salt sits down with a blank sheet of paper and begins crafting a plan without some type of outline. The planning process must follow a logical sequence. A process is defined as a series of actions or steps taken to achieve a particular end. Therefore, regardless of whether your plan is meant to be proactive or reactive in nature, you must still engage in a step-by-step process both to write and carry out your operational objectives.

One of the many beautiful things about the CARVER methodology is its ability to lay concepts out in a logical sequence. When a plan writer begins the process, his/her first step is to analyze the task at hand. Because CARVER is both qualitative and quantitative, this allows the plan writer to use the CARVER acronym to decide how the analysis will unfold. For example, if a security director is assigned to deliver an emergency response plan to his/her organization, he/she can walk through the matrix to ask the questions, prior to putting pen to paper. A sample of the questions that CARVER could spark might be:

C What is ***critical*** about the plan? (why does it need to be written?)

A Who and what will have ***access*** to the plan? (who will use it during an emergency?)

R If there is an emergency, will the plan help the organization ***recover***? (how important is "down time" during an emergency?)

V What are the ***vulnerabilities*** of the plan? (Do factors such as time, money, resources, turnover, communications, interest, etc. affect the plan?)

E What is the ***effect*** of not doing the plan or a minimal version of it? (what is the risk vs. gain of a less-than-comprehensive plan?)

R Who will ***recognize*** the value of the plan? (does the plan need to be written to a specific audience not directly involved with planning, but may be affected as the result of an emergency?)

Obviously, these questions could be asked differently depending on the plan writer's stated requirement. The emphasis here, however, is that just as one might use the time-tested Five W's (who, what, where, when, why) as a starting point, CARVER allows the planner to take the initial planning process one step further.

Variations of the Planning Process Using CARVER

Numerous planning methods exist out there, as well as an exhaustive sampling of plans to be found online. These plans depend on the assets involved, in terms of both assets to be protected and assets that can be utilized as response resources. Because CARVER is primarily focused on asset ranking and categorization, it serves as an ideal methodology to assist in the planning process. For example, if the aim of a business-continuity plan is to maintain operations after disaster strikes, CARVER can be used to address which assets are most necessary for business to continue, at least on a temporary basis. This ranking system not only aids the planner on where to shore up security vulnerabilities prior to an incident, but it also assists him/her in determining what redundancies need to be established (i.e., do spare parts need to be available for certain assets?).

By using a CARVER matrix to determine asset criteria, plans can be drafted to provide responses to emergency incidents. These responses include procedures to maximize safety and minimize property damage, allow essential functions to run uninterrupted for a specific period of time, and communicate critical information to personnel and the public before, during and after a crisis.

In terms of security planning, using the CARVER criteria matrix will mean most of the plans relate to one of the following categories:

- Business Continuity Plans provide guidance and procedures for maintaining essential business operations while recovering from a significant business disruption.

- Business Recovery Plans provide procedures for recovering business operations immediately following a disaster.
- Continuity of Operations Plans provide guidelines and procedures to continue an organization's key essential functions uninterrupted for a specific period of time.
- Crisis Communications Plans provide procedures for disseminating critical emergency communications to personnel and the public before, during and after a crisis.
- Disaster Recovery Plans provide detailed procedures to facilitate recovery of capabilities at an alternate site and is focused on large events.
- Occupant Emergency Response Plans provide planned procedures for maximizing safety and minimizing damage in response to an emergency incident at a facility.

As you can see, the uses for CARVER in response planning are virtually limitless. The CARVER matrix can be constantly reworked to address the question, "What if?" Therefore, the planning process can be examined multiple ways and for multiple types of plans.

CARVER as a Contingency Planning Tool

Contingency planning can anticipate and solve problems that typically arise during the threat assessment and/or an incident response. Therefore, emergency planning and contingency planning are slightly different. The primary difference is the level of specificity of the incident. For example, a preparedness plan can cover several types of weather-related incidents (floods, snow, sandstorms, etc.), while a contingency plan can be highly specific (such as an active shooter incident). Due to its specificity, the contingency-management planning process is organized around four basic components: preparation, analysis, response planning and implementing preparedness. When using the CARVER criteria for contingency planning, you need strong coordination and process management. CARVER can help achieve that management via the following:

- Analyzing potential emergencies
- Analyzing potential impact
- Establishing clear and objective strategies

- Implementing preparedness actions

Effective contingency planning also requires action at several different organizational levels. This action must also be divided between strategic and tactical responsibilities. CARVER can be implemented to score the requirements of each division, unit and individual charged with responding. Typically, these actions include gaining and maintaining situational awareness, activating and deploying key resources and capabilities, effectively coordinating response and, last, demobilizing as the situation permits.

To demonstrate how the CARVER Methodology can be utilized for contingency planning, consider asking the following questions during the planning process, which are also asked during the target analysis and vulnerability assessment:

- What assets are you trying to protect?
- What is your asset inventory?
- What are the risks to these assets?
- Who and what are we defending against?
- How well does the current security solution mitigate these risks?
- Does it work and what are the points-of-failure?
- What other risks does the security solution/response cause?
- Are there ripple-effects to the response?
- What costs and trade-offs does the security solution/response impose or create?
- Is there a cost benefit analysis that can be performed?

By asking the right questions, the planning team can identify specific threats and establish a baseline that can be used against various scenarios to expose vulnerabilities and implement the most appropriate protective measures.

Goal-Setting During the Planning Process

In designing a plan, the planning team must define the agenda, the course of action and/or the target for individuals, the group and the organization. In essence, they must set goals. As part of any successful planning process, the plan should commit to a goal that has been identified as a desired benchmark. The goal should also be publicly stated (or at least identified in the planning document and known to organization members) with an intent to accomplish it.

Note: This is important, because unannounced goals are nothing more than internalized hopes and dreams (this is also important for budgeting, procurement and logistics issues).

Prioritization of these goals is of paramount importance. Goals can be broken down as:

- Critical: goals that are crucial to operational survival
- Enabling: goals that facilitate or enable the plan
- Nice-to-have: goals that are not always necessary but would be useful during an incident

CARVER constitutes a valuable tool for identifying and prioritizing these goals. Though most people in the security industry are familiar with the SMART goal-setting process (Specific, Measurable, Attainable, Resourceful and Timely), using the CARVER matrix can be equally valuable, breaking down the goals into Criticality, Accessibility, Recognizability, Vulnerability, Effect, and Recoverability categories.

Six Steps to Creating Your Plan

Now that we've identified how CARVER can be a useful tool for developing various security plans, it is important to summarize the planning process. For each of the six steps listed below, the CARVER Methodology can be applied to enhance the creation of your organization's desired plan.

Step 1: Develop the first draft and distribute it for input. This includes breaking down the individual courses of action as well as identifying and tagging the necessary resources. It also includes identifying capabilities and performance gaps.

Step 2: Prepare, review and gain approval. Write the plan and approve and disseminate it to the proper reviewers.

Step 3: Refine and execute. Exercise the plan (via tabletop and field exercises) and evaluate its effectiveness. Review, revise and maintain the plan.

Step 4: Assemble a collaborative planning committee. Build the appropriate team.

Step 5: Know your business and understand the situation. Identify targets, hazards and vulnerabilities. Conduct research and analyze the information.

Step 6: Determine and rank your goals and objectives. Clearly identify your operational windows.

By utilizing the CARVER matrix criteria, the planning process can be continually expanded.

Summary

To be able to establish and maintain the appropriate capabilities for managing a range of predictable situations, or the threat of specific emerging crises, you must conduct planning. Plans and planning, although different in scope, are vital to the success of responding to crisis situations. Though many plans and processes currently exist, such measures can be enhanced via the CARVER methodology. Because of CARVER's unique offensive and defensive viewpoints, plan writers and planning teams can better prepare their organizations against adversarial attack and other unforeseen emergencies.

Conclusions and Summary

We will bankrupt ourselves in the vain search for absolute security.

—Dwight D. Eisenhower

Luke and I hope that you found this book interesting. Whether you have taken SMI's CARVER Target Analysis and Vulnerability Assessment Training Course™, or this book is your only introduction to CARVER, our objective was to define CARVER and its versatility. We have used this methodology to conduct vulnerability assessments relating to physical security for a multitude of varied infrastructure, counter-espionage, information technology, red-team operations, strategic business planning, security management consulting processes and targeted intelligence planning. This by no means is a complete list of potential vulnerability-assessment venues using CARVER, but it shows its versatility and capability.

I have provided specific examples of using the CARVER process throughout the book. I have provided some case histories to bring CARVER into the real world. On the next page, I have again provided the table of notional values used in the CARVER physical-security vulnerability assessment to help identify the value matrix used to quantify the CARVER criteria.

The point I want to get across with this table is that the definitions are flexible and fluid. Many of CARVER's values are not set in stone, so please remember that they can be changed to meet the desired assessment objectives. As an example, under Effect (E), the definition for the value 5 is "very high sociological, economic, political impact, considerable loss of life or injuries." In many case, I have modified this statement based on the infrastructure I'm assessing. If it's a school, a single loss of life would most likely qualify as a value 5, because that loss could be a child. However, the statement as written would be accurate in the case of a vehicle-borne-improvised-explosive-device attack against a major shopping mall.

The point I want to make is the versatility that CARVER offers by being creative and making it fit the assessment objectives and the threats.

VALUE	C	A	R	V	E	R
5	Loss would be mission stopper	Easily Accessible. No effective security	Extremely difficult to replace. Long down time	A dedicated adversary hase the capability and expertise to attack	Very high sociological, economic, political impact; cond-ierable loss of life or injuries	Easily recognized by all with no confusion
4	Loss would reduce mission performance considerably	Accessible	Difficult to replace, with long down-time	A dedicated adversary most likely has the capability and expertise to attack	High impact; some loss of life or injuries	Easily recognized by most
3	Loss would reduce mission performance	Somewhat accessible	Can be replaced in a relatively short time	A dedicated adversary most likely does not have the capability and expertise to attack	Moderate impact; some adverse impact on persons	Recognized with some training
2	Loss may reduce mission performance	Difficult to gain access	Easily replaced in a short time	A dedicated adversary most likely does not have the capability and expertise to attack	Little impact; no adverse impact on persons	Hard to recognize. Confusion probable
1	Loss would not affect mission performance	Very difficult to gain access	Immediate replacement. Spare parts are readily available, or asset redundancy	A dedicated adversary does not have the capability and expertise to attack	No unfavorable impact	Extremely difficult to recognize without assistance

I'd also like to take this opportunity to provide an outline of SMI's CARVER Target Analysis and Vulnerability Assessment Training Course™. It is not all-inclusive, however; a detailed outline and agenda can be found at SMI's website: https://www.smiconsultancy.com/the-carver-methodology.html

Module 1: Introduction to Target Analysis and Vulnerability Assessments

- TA/VA course objectives
- Definitions
- Understanding the TA/VA philosophy, benefit/value and complexity

Module 2: CARVER "The Tool to Use"

- History
- Concept
- Process
- Ranking
- Matrix

Module 3: Target Analysis

- What is a target analysis and why is it used and who uses it?
- Methodology and process
- Phases: Statement of requirement, target assessment and operational plan
- Target analysis report

Module 4: Threat Assessment

- Defining adversary capabilities, intent, history, target selection criteria
- Analyzing the threat
- Developing the threat spectrum
- Developing design basis threats
- Developing adversary scenarios

Module 4A: Terrorist Bombs and Bombings

- IED familiarization
- Vehicle bombs

- The suicide bomber
- Suicide bomber operations
- Suicide vehicle bombs

Module 5: Vulnerability Assessments

- Vulnerability assessment process (planning, threat assessment, infrastructure characterization, data analysis, risk-reduction recommendation and reporting phase)

Module 6: Security System Effectiveness

- Define and implement; deterrence, detection, delay and response and, if required, development of a mitigation plan

Module 6A: Perimeter Inspections

- Perimeter systems
- Fences and barriers
- Gates
- Access control
- Perimeter sensors

Module 6B: Building Security

- Building types of threats
- Access control
- Doors
- Windows
- Nontraditional access points
- Access control hardware and systems

The above course agenda is not complete. There are many sub-elements in each of the bullet points which drill down into the details of each module's content. Again, additional information on course content can be found on SMI's website.

Again, Luke and I would like to thank you for taking an interest in this book, and we hope it meets some of your security requirements. If you have any questions, we can be reached through email or phone. Contact information is located on the SMI website at www.smiconsultancy.com.

I also want to take this opportunity to provide the reader with templates that can be used in crafting threat assessments and final reports. You can find them in Appendices I-III. I hope they will be useful, because in some cases writing threat assessments can be most tedious. You will need to conduct a lot of research to compile the information necessary for threat and vulnerability assessments. I hope the two case studies I provided – covering the international convention center and sports stadium, and the energy infrastructure, will lessen the difficulty. As you will see, much of the information in the appendices was covered in previous chapters. Nevertheless, they should become valuable resources.

The world is still a dangerous place. When I created the CARVER methodology over 40 years ago with my colleagues from the CIA, I never could have imagined I would still be using it all these years later. I guess that means it is still relevant. May it serve you and keep you safe.

—Leo

Appendix I

Sports Venue Threat Assessment Template

CONTENTS

Convention Center and Sports Stadium Design Basis Threats
Introduction
The elements of a DBT

Introduction

This threat assessment is largely based on open source information found on the internet and documents authored by the Department of Homeland Security (DHS), titled "Threats to College Sports and Entertainment Venues and Surrounding Areas" and the Texas State University, titled "United States Active Shooter Events from 2000 to 2010: Training and Equipment Implications". Additionally, we have provided information on gang related activities surrounding the Convention Center and Sports Stadium which was obtained through open source and interviews with Law Enforcement in the Multi Agency Gang Task Force (M.A.G.T.F.) and in discussions with County Sheriff's Deputies that patrol the area of the Convention Center and Sports Stadium. That information is incorporated into the body of this assessment where applicable.

The overall threat to the Convention Center and Sports Stadium facility is considered LOW, but the overall threat to events is considered MODERATE to HIGH depending on the type of event it is hosting.[3]

> **"Hundreds of U.S. college sports and entertainment venues, including indoor arenas and outdoor stadiums, attract large numbers of spectators. Infrastructure around the venues and campus complexes, including hotels, local businesses, and parking areas, often attract large crowds before and after an event. Some events are particularly high profile, attracting national media attention and crowds in excess of 100,000".**

High threat events at the Convention Center and Sports Stadium include, but are not limited to:

- Sporting Events
- Concerts
- Trade Shows

---

[3] Department of Homeland Security, Office of Intelligence and Analysis, "Threats to College Sports and Entertainment Venues and Surrounding Areas"

- Religious Services & Conventions
- Ethnic Festivals
- Political Rallies
- Other Corporate, Community and University Events.

The threats to the Convention Center and Sports Stadium should be considered a snapshot in time, as a specific event can increase or decrease the threat.

Background

Historically, those who have been responsible for protecting our nation's critical infrastructure and other vital facilities have presumed that a major attack on them would most likely occur in cities near our borders with Canada or Mexico (such as Seattle, San Diego and El Paso), which offer greater ease of access; or in large metropolitan areas (such as New York City, Washington D.C., Los Angeles, San Francisco and Boston), where the notoriety and impact of an attack would be greatly amplified.

Since the destruction of the Murrah Federal Building in Oklahoma City on April 19, 1995, the terrorist attacks of September 11, 2001, and other recent tragedies such as the Boston Marathon Bombing, these presumptions have, for the most part, been set aside. There are several reasons for this. First, it has become obvious to many observers that terrorists and other adversaries can and will attack targets wherever the opportunity arises. Typically, they look for easy targets that are unsecured and manifest exploitable weaknesses. As larger cities and urban areas, often with Federal and other financial assistance, take meaningful steps to eliminate their vulnerabilities and improve their security, the attractiveness of unprotected targets grows considerably. Today, regardless of size or location, the threat of an attack is greatest where the defenses are weak and the attention to security is lax.

A vital first step in conducting a vulnerability assessment for a fixed asset, such as a sports or entertainment venue, is to perform a threat assessment. In performing a threat assessment, we assume that the asset is a target for attack. The term "target" means the asset has potential adversaries both inside and outside of its established boundaries. The term "attack" means any action intended to bring about harm, violence, damage or destruction. The primary goal of a threat assessment is to determine the likely threats against a given target based upon all available information.

The information gathered in this threat assessment is used in the vulnerability assessment to assess an asset's vulnerability to potential threats, the effectiveness of existing security measures against such threats, and the changes that are necessary to minimize the threats and/or the consequences thereof. The information in this threat assessment is also used to develop design basis threats (DBTs) appropriate to an asset.

In order to develop the Threat Spectrum of likely threats to the Convention Center and Sports Stadium we have to characterize the potential adversaries. The adversary and threat will vary as to targeting the facility itself or an event at the facility. Their method of attack will vary greatly based on their capabilities. Some examples of potential adversaries:

- International and Domestic Terrorist Organizations
- Hate groups, militias, cults, radical environmentalists and opposition movements
- Vandals and computer hackers
- Disgruntled employees, contractors and other insiders
- Disgruntled customers
- Criminals
- Gang Activity

This threat assessment is divided into sections according to the goals and objectives of the vulnerability assessment:

Potential Adversaries

- What groups or persons are most likely to pose a threat to the target?
- Why are these groups or persons threats?
- What types of attacks have occurred in the past, and may happen again?
- What are the objectives or motivations of the potential aggressors?

Capabilities

- What are the capabilities of the potential adversaries?
- What is their access to hazardous items?

Mode of Attack

- How are the threats likely to be carried out?
- What is the likelihood that the threats will be carried out?

Potential Consequences

- What are the likely effects of an attack?
- What is the expected response time?

Each of these groups are discussed below as to the threat they pose to the Convention Center and Sports Stadium:

Threat Assessment

The catastrophic events of September 11th were a milestone in international terrorism, followed by the Madrid train bombings, London train and bus bombings, and the attacks in Mumbai. The planning, execution and magnitude of the attacks exceeded pre-9/11 attacks anywhere in the world. Global terrorism has become an unbounded threat utilizing more aggressive, horrific and deadly tactics. Domestic terrorism has equally become an unbounded threat as characterized by the Ft Hood, Century Theater, Sandy Hook School and other active-shooter incidents and the Times Square attempted bombing, and Boston Marathon Bombing, etc.

This is not to say that the Convention Center and Sports Stadium must prepare for a major terrorist attack, but the potential of such an attack cannot be ruled out.

Hate groups, militias, cults, and opposition movements that are intent on violent action can also pose a serious threat to Convention Center and Sports Stadium events. Such organizations often use tactics and methods of attack that are similar to, and largely indistinguishable, from those used by terrorists. Activist groups and cults are often intent on making a public political statement. While such groups may not be state sponsored or have ties to foreign governments, they often harbor a hatred of the United States, the Federal and local governments, large organizations and business enterprises, and other symbols of power or authority.

> "DHS/I&A, the SCIIC, and the MIOC lack credible or specific information indicating current plots aimed at sports and entertainment venues but assess that terrorist organizations such as al-Qaida, its affiliates, and ideologically similar groups view occupied sports stadiums and arenas as potential targets. Al-Qaida's repeatedly stated objectives include causing economic damage, mass casualties, and psychological impact, all of which could be achieved by attacking a large sporting or entertainment event.
>
> A posting on a jihadist message board in February 2006, "How You Can Kill Thousands of Americans with a Few Hundred Dollars and Three Men," advocated targeting U.S. sporting venues and other public facilities during events. The posting did not describe or imply an imminent attack.
>
> As an example, Six al-Qaida-inspired men arrested for plotting to attack Fort Dix also allegedly considered targeting the annual Army-Navy football game.
>
> Because of the heavy security presence at many college venues, plotters may also consider attacking surrounding targets such as public transportation, large gatherings, or businesses. Terrorists may be able in some cases to achieve similar impact without having to defeat stadium security."

There are no indications of any domestic terrorist threats against sports venues, as they tend to limit their interest to a fairly narrow spectrum of sites and institutions, usually those antithetical to their cause. A domestic extremist may, however, choose to attack a college's sports or entertainment venue in response to a grievance specific to the targeted institution or event.

Lone Wolf Threat

The Lone Wolf is considered one of the greatest threats we face in this climate of world turmoil. The Lone Wolf can be a domestic or international extremist with no known ties to an extremist group or groups, a psychotic individual, or any other person bent on committing an atrocity. The following assessment is provided by DHS [5]:

---

[4] Department of Homeland Security, Office of Intelligence and Analysis, "Threats to College Sports and Entertainment Venues and Surrounding Areas"

"College sporting events are potentially threatened by lone wolves, so called because they operate individually, with little or no outside assistance or guidance, and can be motivated by wide-ranging factors. Because of their nature and lack of obvious ties to known groups of concern, lone wolves present a particular challenge for law enforcement.

A heavily armed individual, upset at a local decision to deny his business a liquor license, drove to the site of the 2008 Super Bowl on game day. The subject intended to go on a pre-game shooting rampage, but changed his mind at the last moment and turned himself in to police.

In October 2005, a University of Oklahoma student blew himself up in a park outside the Gaylord Family-Oklahoma Memorial Stadium while a game was in progress. Authorities found no ties to terrorism, but the incident's proximity to the stadium was cause for alarm."

Insider Threat - Disgruntled Employees, Contractors, and Other Insiders

Insiders include present and former employees, contractors, vendors, independent consultants, service providers and other persons who have access to or work within the boundaries of a given asset. Insiders who have authorized access to a facility or asset typically pose the greatest threat to it because they have knowledge of its operations, vulnerabilities and security systems. Insiders also pose a significant threat because access controls and other security systems that are intended to deter, detect and delay outside intruders are often ineffective against insiders. The threat posed by insiders is further exacerbated by the fact that such persons are often times not subject to any meaningful identity checks or background investigations prior to their employment or retention. Moreover, poorly vetted non-employee janitors, groundskeepers, maintenance personnel and utility service technicians are often given unrestricted and unescorted access to sensitive areas and equipment that even trusted senior employees do not have.

Insiders are generally motivated to action for reasons of anger and revenge (especially in the case of disgruntled employees) or

[5] Department of Homeland Security, Office of Intelligence and Analysis, "Threats to College Sports and Entertainment Venues and Surrounding Areas"

money. Terrorists and activist group members may attempt to circumvent existing security systems by seeking to become an insider through employment within the system or by recruiting and employee to gain access. The actions and motivations of such persons, however, are fundamentally different than those of a true insider. It is necessary, then, to differentiate between "masquerading" insiders from "true" insiders. The former seek employment with the intent of committing harm, and the latter enter into employment without intent to do harm but gradually develop the desire over time.

The following assessment is provided by DHS :

> **"Attack planners could attempt to gain entry to a secure college sports venue by leveraging the access of an employee or by impersonating an authorized individual. Counterfeit law enforcement badges and credentials purchased online—often of high quality and hard to differentiate from legitimate credentials—can facilitate unauthorized entry. Use of illicit credentials to gain access to the grounds of the sporting events may be difficult to detect.**
>
> **In 2005, an individual impersonating an FBI Special Agent was able to enter secure areas of the track venue during a professional stock car race. Security personnel or staff did not question his presence, and he was able to move freely within restricted areas."**

Vandals and Computer Hackers

Vandals and computer hackers usually commit crimes of opportunity – illegal actions performed without provocation, significant premeditation or elaborate planning. Individuals who dump contaminants into generator gas tanks, or who deface or damage field equipment, or post protest messages on an organization's website, are typical examples of this type of threat. Vandals and computer hackers often act on immediate urges for their own personal enjoyment and satisfaction, or to impress peers. Typically, their actions and the consequences of them are not well-considered, and these persons are often indifferent to or unaware of the harm that they might cause or have caused. They often act simply because they

[6] Department of Homeland Security, Office of Intelligence and Analysis, "Threats to College Sports and Entertainment Venues and Surrounding Areas"

have the means to do so, and believe that they will not be caught. However, the unplanned, unstructured and rash nature of these types of attacks means that they are far from harmless.

Computer hacking can be an electronic form of vandalism. In many cases, the computer hacker seeks to access an IT network, system, database or website, with the goal of defacing, destroying, disabling, damaging or seizing control of it. However, with computer-based attacks, you can never be certain whether the intruder is a computer hacker randomly probing for vulnerabilities, or a determined terrorist testing the security of a specific system.

Gang Activity

Gang activity for the local area is growing. This includes the gang population in all of the surrounding cities.

According to the Federal Bureau of Investigation's 2009 National Threat Assessment, there are over one million gang members nationwide. This number was expected to increase 25,000 every year past 2009. The FBI also states in the same threat assessment that up to 80% of all violent crimes can be linked to gangs and gang members.

It is for this reason that understanding the gang epidemic in any area is vital when trying to conduct a threat assessment. For this reason, an analysis of area gangs and similar type criminal activity was conducted of the area surrounding the Convention Center and Sports Stadium. This analysis was completed by obtaining open source intelligence (including blogs, tips and notes) from social networking sites about gangs and gang activity in the area, interviewing Law Enforcement and the Multi Agency Gang Task Force (M.A.G.T.F.) and speaking with County Sheriff's Deputies that patrol the area of the Convention Center and Sports Stadium. Moreover, investigators toured the area surrounding the Convention Center and Sports Stadium both day and night time. The tour of the surrounding area also included observing "The Mall", (a mall named for its close proximity to Convention Center and Sports Stadium) which is located 1.3 miles from the Convention Center and Sports Stadium.

[7] Federal Bureau of Investigation - 2009 National Threat Assessment

Upon compiling all of the information and statistical data, what was learned raised some concerns. In the area surrounding the Convention Center and Sports Stadium, no less than 10 criminal street gangs claimed territory or "Turf". This turf either bordered on the Convention Center and Sports Stadium or was within walking distance from Convention Center and Sports Stadium grounds. Within the 10 gangs in the area, there are 27 "sets" or subgroups. Of the gangs found, there are 1,348 documented criminal street gang members in the County. Additionally, within the last calendar year from September 2012 to August 2013, there were no less than 2,000 crimes of violence (statistics reported from the County Sheriff's Office), in the immediate area surrounding the Convention Center and Sports Stadium. If the F.B.I. is correct in the figures given out in the National Street Gang Assessment, then 1,600 violent crimes were committed by criminal street gang members within the last calendar year in the geographic area surrounding the Convention Center and Sports Stadium. Some might say that gang activity in the area surveyed is currently rampant.

Gangs are well-versed in numerous ways of obtaining illegal monies. In the Convention Center and Sports Stadium setting, the "number one" way for any gang to thrive is through the sale of illegal narcotics. Narcotics are just one way for any gang to try to make money. In this type of environment, everyone is the equivalent of walking ATM machines to gang members. Other crimes committed are robberies and burglaries. There were over 150 robberies in the District area within the last calendar year. These robberies consisted of carjacking, home invasions and simple person on person crime. It is noted that, while this report was being written, County Sheriff's Office, Police Department and Convention Center and Sports Stadium police were currently in a man hunt for a serial rapist who had committed four home invasions in one night and several more in the past. There is no word yet on the determination of whether this person was a gang member or possible gang associate.

When asking why crime in the area of the Convention Center and Sports Stadium is so high, it is important to try to determine what gang members see in staying in the area, and why they won't move on to another more profitable location. The reasoning is simple. There is no other area more profitable for a gang member than the area surrounding the Convention Center and Sports Stadium.

Sporting events also appeal to area gang members. This is in part because numerous gang-related crimes that occur at sporting events go unreported. Ticket scalping, car burglaries, pickpocketing, robberies, illegal alcohol sales and narcotics sales are just the tip of the iceberg of crimes that occur at any sporting event, be it collegiate or professional. Outside of law enforcement and the gang world, there is a well-known fact that for every professional and/or college sports team, a gang somewhere in the nation will use that team's sports jerseys and paraphernalia to represent their gang. Area law enforcement have known for years that gangs use sports clothing for this purpose and have documented gang members under State Statute in some cases (in part) for the clothing that gang members wear.

It is for these and other reasons that area gangs are so well entrenched in the area surrounding the Convention Center and Sports Stadium, and why the gang problem is such a difficult one to solve. This high crime area breeds gangs and their numbers for both financial gain and expansion/protection of the area crime market.

Capabilities

The capabilities of potential adversaries vary from one type of group to the next.

Terrorist Organizations

Many terrorist groups are well-organized, well-trained and well-funded, and their members are experienced, dedicated and prepared. They conduct extensive research on their targets, identify their weaknesses, and plan their attacks. We know that members of Al Qaeda, other terrorist groups and the lone wolf extremist often receive training in the use or conduct of:

- Firearms, explosives and incendiary devices
- Biological agents, chemical substances, hazardous materials, and other contaminants

- Hand tools, small power tools and other similar equipment
- False identities, disguises and clandestine operations
- Reconnaissance, casing, surveillance, photography and observation

Hate Groups, Militias, Cults, Radical Environmentalists and Opposition Movements

While some activist groups and cults are well-organized, most lack the organization, training, and funding that terrorist groups usually possess. As demonstrated by the recent worldwide protests of meetings of the International Monetary Fund and the World Bank, the Internet has made it possible for even unaffiliated opposition movements to identify mutual targets of opportunity, coordinate their activities, plan joint or concurrent events, collaborate on common goals, and share information on tactics. Members of such groups sometimes receive training in:

- Firearms, explosives and incendiary devices
- Biological agents, chemical substances, hazardous materials, and other contaminants
- Non-violent protests and resisting arrest
- Sabotage and harassment

Vandals and Computer Hackers

Vandals normally do not have any special capabilities or training, and are likely to use whatever means and materials are readily available. Computer hackers, on the other hand, have extensive knowledge of IT systems, computer networks, system vulnerabilities, attack methodologies, malicious software, and IT security protocols. They have at their disposal and use a wide range of tools that assist them in probing systems, discovering exploitable weaknesses, launching attacks, and making it difficult for others to uncover and trace their actions. Hackers are also known to share technical and other information on system exploits over the internet.

Disgruntled Employees, Contractors, and Other Insiders

Insiders often possess one or more methods of accessing their targets (e.g., keys, combinations, passwords) that, when coupled with stolen confidential non-public information, can present a very real risk of undetected illegal activity within the infrastructure. In many cases, employee and consultant insiders also have specialized technical or operating knowledge of relevant machinery, equipment, facilities, processes, procedures, access controls, security mechanisms and protocols, and countermeasures.

Gangs

Local gangs have extensive capability to commit criminal activities in and around the Convention Center and Sports Stadium. Based on the threat assessment of gangs and related criminal activity in the area surrounding the Convention Center and Sports Stadium, it can be assumed that their activities will eventually affect the Convention Center and Sports Stadium.

Mode of Attack – The Threat Spectrum

An attack against the Convention Center and Sports Stadium could take one of the following forms:

- The Physical Destruction of the facility or asset
- The release of Hazardous Material (HAZMAT) during an event
- The theft of property from the Convention Center and Sports Stadium facility, parking lots, loading docks, etc.
- Improvised Explosive Device (IED) or Vehicle-Borne Improvised Explosive Device (VBIED)
- Active Shooter Event (ASE) with the goal of multiple casualties
- Vandalism of the facility or asset to include defacement and destruction of property

Threat Spectrum

If we take each of these potential modes of attacks and plot them on a graph as to their Impact and Likelihood, we can see that the most likely events are Theft and Vandalism, although they have the least impact on facility operations. That being said, it is imperative that the Convention Center and Sports Stadium protect its facility, particularly when hosting events, against the high-impact types of attacks, which would likely cause death and injuries.

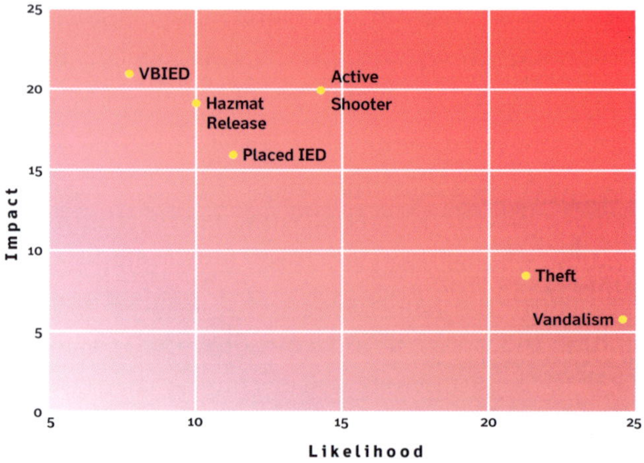

Potential Consequences

Based on the threat spectrum, the Convention Center and Sports Stadium is subject to various consequences resulting from the mode of attack and stadium occupancy. The consequence is heightened during events where large numbers of people are in attendance and a mode of attack is employed to kill or injure. Although these modes of attack are less likely than theft or vandalism, the penalties are high if successful.

Conclusion

The available information and recent experiences suggest that sports arenas and events could be the target of a variety of adversaries. Terrorists and other adversaries seek out unprotected or "soft" targets; unfortunately, most of America's sports venues currently qualify as such. Facilities that fail to take proactive measures to evaluate their vulnerabilities, shore up their security systems, implement threat detection measures, update their emergency operations plans, and train their employees to handle incidents, will find that the likelihood of an attack will remain.

Convention Center & Sports Stadium Design Basis Threats

Introduction

In order to design and evaluate security systems and programs that will effectively protect critical infrastructure, the methods and capabilities of potential adversaries must first be understood and defined. Using the information developed in the threat assessment, and applying our experience in conducting vulnerability assessments for a wide range of assets, we have provided the following design basis threats (DBTs) as some of the potential scenarios which might affect the Convention Center and Sports Stadium. According to the U.S. Government, Design Basis Threat is a profile of the type, composition and capabilities of an adversary.

The elements of a DBT incorporate the following attributes of a perceived threat:

- Adversary Description and Capability
- Type of Equipment
- Knowledge
- Tactics/MO
- Goals

Although every threat cannot be predicted, the security team has listed the most likely DBTs according to the intelligence collected. The Convention Center and Sports Stadium Asset Risks and Probability of Attack (Pa) were based on the following design basis threats:

Outsider(s) - Domestic/International Terrorist(s) — Vehicle Borne Improvised Explosive Device

An attack planned by 2-3 outsiders with access to, knowledge of, and/or experience with explosives. They will use sophisticated hand tools, power tools, firearms and have the other resources necessary to obtain and/or make a vehicle borne improvised explosive device (VBIED). Their goal is to utilize a VBIED against the Convention Center and Sports Stadium, during an event, so as to kill and injure numerous persons and cause significant destruction of the facility infrastructure, denying its use for a lengthy period of time thereby causing serious economic consequences for the Convention Center and Sports Stadium. The attackers are not opposed to a suicide attack and will harm anyone hindering their efforts.

Outsider - Domestic/International Terrorist – Placed Improvised Explosive Device

An attack planned by 2-3 outsiders with access to, knowledge of, and/or experience with explosives. The attackers will use sophisticated hand tools, small power tools and firearms and will be trained in the use of explosives and/or incendiary devices. Their goal is to utilize one or more IEDs placed in the Convention Center and Sports Stadium during or before an event, so as to kill and injure numerous persons. The attackers are not opposed to a suicide attack and will harm anyone hindering their efforts.

Outsider - Domestic/International Terrorist – HAZMAT

An attack planned and conceived by 2-3 outsiders without authorized access, but with knowledge of the Convention Center and Sports Stadium and its air conditioning system obtained through open source information, observation and solicitation. The attackers will use sophisticated hand tools, small power tools and firearms and will be trained in the use of hazardous materials. Their goal is to vent chlorine or other hazardous gas into the air intake system during an event to cause panic and inflict casualties in the arena. They will harm anyone hindering their efforts.

Outsider – Active Shooter Event

An attack planned by a single individual with knowledge of the Convention Center and Sports Stadium from attending prior events. The attacker is well aware of the numerous access points where undetected entry can be accomplished. The attacker will enter the Convention Center and Sports Stadium during an event with multiple weapons and ammunition. The goal is to kill or wound as many attendees as possible before law enforcement can intervene. They will harm or kill anyone hindering their efforts.

Outsider - Vandal

An attack planned and conceived by a single or multiple outsider(s) without authorized access or inside information using unsophisticated hand tools or small power tools with the intent of defacing or damaging assets of the Convention Center and Sports Stadium. The outsider does not intend to cause physical harm to staff or attendees.

Outsider - Theft

An attack planned and conceived by a single or multiple outsider(s) without authorized access or inside information using unsophisticated hand tools or small power tools with the intent of stealing assets of the Convention Center and Sports Stadium and/or its patrons and staff. Potential targets would be the loading dock, offices, lockers, parked cars, food storage areas, equipment, etc. The outsider does not intend to cause physical harm to staff or attendees.

Insider – Theft

An attack planned and conceived by a single or multiple insider(s) with authorized access or inside information using unsophisticated hand tools or small power tools with the intent of stealing assets of the Convention Center and Sports Stadium, and/or its patrons and staff. Potential targets would be the loading dock, offices, lockers, parked cars, food storage areas, equipment, etc. The insider does not intend to cause physical harm to staff or attendees.

Appendix II

Energy Sector Threat Assessment Template

The following threat assessment is provided as a template that can be used for many different types of infrastructure such as, energy, water, sporting venues, buildings etc.

CONTENTS

Threat Overview

The importance of energy security and energy infrastructure security cannot be overstated. It is among the most serious security, economic and environmental challenges today and is expected to continue to be such in the future.

The energy sector provides the fuel that keeps the global economy moving and our societies working. The disruption or destruction of these infrastructures would have a serious impact on the health, safety, security and economic well-being of individuals and the world as a whole.

The following threat assessment is largely based on open source information, unclassified reports received from intelligence agencies, FBI and law enforcement, as well as documents authored by the following agencies:

- Department of Energy (DOE)
- North American Electric Reliability Council, (NERC)
- Sandia National Laboratories, Threat Analysis Framework
- International Atomic Energy Agency
- National Research Council of the National Academies
- Organization for Security and Co-operation in Europe

A terrorist attack on the power system would lack the dramatic impact of the attacks in New York, Madrid, or London. It would not immediately kill many people or make for spectacular television footage of bloody destruction. But if it were carried out in a carefully planned way, by people who knew what they were doing, it could deny large regions of the country access to bulk system power for weeks or even months. An event of this magnitude and duration would lead to turmoil, widespread public fear, and an image of helplessness that would play directly into the hands of the adversary(s). If such large extended outages were to occur during times of extreme weather, they could also result in hundreds or even thousands of deaths due to heat stress or extended exposure to extreme cold.

The U.S. Power Grid is enormously complex. There are over 11,000 generation facilities, in excess of 200,000 miles of high voltage transmission lines and thousands of substations that constitute one of the largest machines ever constructed in the United States.

The grid, consisting of a network of substations, transmission lines, distribution lines and other components can easily be targeted by those intent on attacking our critical infrastructure. Today most power is generated by large central generating stations that are located far from local law enforcement agencies that protect them. The power grid is an attractive target because most of its assets are geographically dispersed and hence very vulnerable. Since a high percentage of electrical assets operate in remote locations, it becomes difficult, if not impossible, to defend against explosives or projectiles fired from a distance.

The power grid also includes less visible assets that are used to run the system such as, Supervisory Control and Data Acquisition (SCADA) systems, communication systems, and the scores of people that are needed to operate them effectively.

The overall threat to energy facilities is considered MODERATE to LOW. However, threats to the energy sector have been increasing, and most experts believe it is only a matter of time before a major attack is conducted directly against an electric utility or indirectly against the community it serves. Due to critical dependencies and interdependencies within the energy sector, the loss of electric power due to an attack, could quickly cascade to other lifeline infrastructure sectors, such as water, wastewater, communications, transportation, and information technology, potentially degrading essential services necessary for public health and safety. It is clear that large scale or long duration power outages can easily result in economic costs of billions of dollars. A systematic and coordinated terrorist attack on the power grid could result in considerably more damage than we have previously experienced from storms or natural disasters.

The threats to energy facilities should be considered a snapshot in time, as a specific domestic or international event can increase or decrease the threat.

Background

Historically, those who have been responsible for protecting our nation's critical infrastructure and other vital facilities have presumed that a major attack on them would most likely occur in cities near our borders with Canada or Mexico such as Seattle, San Diego and El Paso which offer greater ease of access. Or in large metropolitan areas such as New York City, Washington D.C., Los Angeles, San Francisco and Boston where the notoriety and impact of an attack would be greatly amplified.

Since the destruction of the Murrah Federal Building in Oklahoma City on April 19, 1995, the terrorist attacks of September 11, 2001, and other recent tragedies such as the Boston Marathon Bombing, these presumptions have, for the most part, been set aside. There are several reasons for this. First, it has become obvious to many observers that terrorists and other adversaries can and will attack targets wherever the opportunity arises. Typically, they look for easy targets that are unsecured and manifest exploitable weaknesses. As larger cities and urban areas, often with Federal and other financial assistance, take meaningful steps to eliminate their vulnerabilities and improve their security, the attractiveness of unprotected targets grows considerably. Today, regardless of size or location, the threat of an attack is greatest where the defenses are weak and the attention to security is lax.

A vital first step in conducting a vulnerability assessment for the energy sector's fixed assets is to perform a threat assessment. In performing a threat assessment, we assume that the asset is a target for attack. The term "target" means the asset has potential adversaries both inside and outside of its established boundaries. The term "attack" means any action intended to bring about harm, violence, damage or destruction. The primary goal of a threat assessment is to determine the likely threats against a given target based upon all available information.

The information gathered in this threat assessment is used in the vulnerability assessment to evaluate an asset's vulnerability to potential threats, the effectiveness of existing security measures against such threats, and the changes that are necessary to minimize the threats and/or the consequences thereof. The information in this threat assessment is also used to develop design basis threats (DBTs) and a Threat Spectrum appropriate to an asset(s).

> *A fundamental principle of physical protection is that it should be based on the State's current evaluation of the threat. This evaluation is formalized through a threat-assessment process. A DBT is derived from this threat assessment. To define the DBT, the set of threats described in the State's Threat Assessment are refined to take account of other*

issues (for example technical, economic, and political issues) and the particular requirements of planning for system design. To make the transformation from threat assessment to DBT, rigorous analysis and decision-making are essential

—International Atomic Energy Agency

In order to develop the Threat Spectrum of likely threats to the energy sector, we have to characterize the potential adversaries. The adversary and threat will vary as to targeting the facility itself or a remote asset of the facility. Their method of attack will vary greatly based on their capabilities. Some examples of potential energy sector adversaries follow:

- International and domestic terrorist organizations
- Hate groups, militias, cults, radical environmentalists and opposition movements
- Vandals and computer hackers
- Disgruntled employees, contractors and other insiders
- Disgruntled customers
- Criminals

This threat assessment is divided into sections according to the goals and objectives of the vulnerability assessment:

Potential Adversaries

- What groups or persons are most likely to pose a threat to the target?
- Why are these groups or persons a threat?
- What types of attacks have occurred in the past, and may happen again?
- What are the objectives or motivations of the potential aggressors?

Capabilities

- What are the capabilities of the potential adversaries?
- What is their access to hazardous items?

Mode of Attack

- How are the threats likely to be carried out?
- What is the likelihood that the threats will be carried out?

Potential Consequences

- What are the likely effects of an attack?
- What is the expected response time?

Each of these groups are discussed below as to the threat they pose to the energy sector.

International Terrorist Threat

The catastrophic events of September 11th were a milestone in international terrorism, followed by the Madrid train bombings, London train and bus bombings, and the attacks in Mumbai and most recently, the attacks in Paris, France, and Brussels, Belgium. The planning, execution and magnitude of the attacks exceeded pre-9/11 attacks anywhere in the world. Global terrorism has become an unbounded threat utilizing more aggressive, horrific and deadly tactics. Domestic terrorism has equally become an unbounded threat as characterized by the Fort Hood, Century Theater, Sandy Hook School and other active-shooter incidents, as well as the Times Square attempted bombing, and Boston Marathon bombing, etc.

This is not to say that the energy sector must prepare for a major terrorist attack, but the potential of such an attack cannot be ruled out.

The power grid can represent an attractive target for terrorists who seek to inflict not only physical but also economic or environmental damage. Researchers indeed suggest that the threat of attacks by terrorists and other non-state actors against critical infrastructure, especially in the field of cyber/IT security, is growing.

As an example of the domestic/international threat against utilities, members of the drug cartels in Michoacán, Mexico, armed with guns and Molotov cocktails, attack six fuel stations and several power plants in the state, leaving some 420,000 people without electricity for hours. "They were acts of terrorism" that targeted "strategic services of the federation," Vallejo said.

The governor said the authorities had heard about "this possible scenario" a week ago but "not of this magnitude" and that federal police and soldiers were prepared for it. The sabotage occurred despite the presence of troops deployed by the federal government in May to bring peace to the state, which has endured violence perpetrated by the Knights Templar drug cartel.

Domestic Terrorist Threat

Hate groups, militias, cults, environmental extremists, and opposition movements that are intent on violent action can also pose a serious threat to the energy sector. Such organizations often use tactics and methods of attack that are similar to, and largely indistinguishable from those used by terrorists. Activist groups and cults are often intent on making a public political statement. While such groups may not be state sponsored or have ties to foreign governments, they often harbor a hatred of the United States, the Federal and local governments, large organizations and business enterprises, and other symbols of power or authority.

In response to new gun-control laws passed in New York in January 2013, a militia extremist posted an article on line that provided examples of how the government could lose a civil conflict. The article showed how to attack the power grid and highlighted vulnerabilities that could easily be exploited by lone wolves or small teams. http://www.bob-owens.com/2013/01/shock-the-system/

Lone Wolf Threat

The Lone Wolf is considered one of the greatest threats we face in this climate of world turmoil. The Lone Wolf can be a domestic or international extremist with no known ties to an extremist group or groups, a psychotic individual, or any other person bent on committing an atrocity. As an example, more than 10,000 people in Arkansas were dumped into a blackout Sunday, October 10, 2013, following an attack on that state's electric grid, the third such attack in recent weeks. In August, a major transmission line in the region around Cabot, AR, was deliberately cut. According to the FBI, in the early morning hours of September 29, 2013, officials with Entergy Arkansas reported a fire at its Keo substation located on Arkansas Highway 165 between Scott and England in Lonoke County. Fortunately, there were no injuries and no reported power outages. Investigation has determined that the fire, which consumed the control house at the substation, was intentionally set. The person or persons responsible for this incident inscribed a message on a metal control panel outside the substation which reads, "YOU SHOULD HAVE EXPECTED US"

The FBI and Joint Terrorism Task Force (JTTF) arrested a 37-year-old Arkansas man in connection with the series of attacks on the local power grid. The suspect, Jason Woodring, is a resident of Jacksonville, Arkansas. Woodring was charged with destruction of an energy facility. If convicted, he could face a prison sentence of up to 20 years imprisonment.

SAN JOSE, CA—the attack began just before 1 a.m. on April 16, 2013 when someone slipped into an underground vault not far from a busy freeway and cut telephone cables. Within half an hour, snipers opened fire on a nearby electrical substation. Shooting for 19 minutes, they surgically knocked out 17 giant transformers that funnel power to Silicon Valley. A minute before a police car arrived, the shooters disappeared into the night. Nobody has been arrested or charged in the attack at PG&E Corp.'s Metcalf transmission substation. It is an incident of which few Americans are aware. But one former federal regulator is calling it a terrorist act that, if it were widely replicated across the country, could take down the U.S. electric grid and black out much of the country.

Insider Threat - Disgruntled Employees, Contractors, and Other Insiders

Insiders include present and former employees, contractors, vendors, independent consultants, service providers and other persons who have access to, or work within the boundaries of a given facility or asset. Insiders who have authorized access to a facility or asset typically pose the greatest threat to it because they have knowledge of its operations, vulnerabilities and security systems. Insiders also pose a significant threat because access controls and other security systems that are intended to deter, detect and delay outside intruders are often ineffective against insiders. The threat posed by insiders is further exacerbated by the fact that such persons are often times not subject to any meaningful identity checks or background investigations prior to their employment or retention. Moreover, poorly vetted non-employee janitors, groundskeepers, maintenance personnel and utility service technicians are often given unrestricted and unescorted access to sensitive areas and equipment that even trusted senior employees do not have.

Insiders are generally motivated to action for reasons of anger and revenge (especially in the case of disgruntled employees) or money. Terrorists and activist group members may attempt to circumvent existing security systems by seeking to become an insider through employment within the system or by recruiting an employee to gain access. The actions and motivations of such persons, however, are fundamentally different than those of a true insider. It is necessary, then, to differentiate between "masquerading" insiders from "true" insiders. The former seek employment with the intent of committing harm, and the latter enter into employment without intent to do harm but gradually develop the desire over time.

Vandals and Computer Hackers

Vandals and computer hackers usually commit crimes of opportunity – illegal actions performed without provocation, significant premeditation or elaborate planning. Individuals who dump contaminants into generator gas tanks, or who deface or damage field equipment, or post protest messages on an organization's website, are typical examples of this type of threat. Vandals and computer hackers often act on immediate urges for their own personal enjoyment and satisfaction, or to impress peers. Typically, their actions and the consequences of them are not well-considered, and these persons are often indifferent to or unaware of the harm that they might cause or have caused. They often act simply because they have the means to do so, and believe that they will not be caught. However, the unplanned, unstructured and rash nature of these types of attacks means that they are far from harmless.

Computer hacking can be an electronic form of vandalism. In many cases, the computer hacker seeks to access an IT network, system, database or website, with the goal of defacing, destroying, disabling, damaging or seizing control of it. However, with computer-based attacks, you can never be certain whether the intruder is a computer hacker randomly probing for vulnerabilities, or a determined terrorist testing the security of a specific system.

Due to SCADA control systems physically being disconnected from the internet (referred to as an air-gap); utilities often assume that bad things like hackers and worms can never get into their critical control systems. However, recent concerns exist over the potential for hackers to gain physical access into SCADA systems through remote locations. These remote locations often have cyber access points that could be exploited by hackers in order to gain access to critical cyber assets.

Capabilities

The capabilities of potential adversaries vary from one type of group to the next.

Terrorist Organizations

Many terrorist groups are well-organized, well-trained and well-funded, and their members are experienced, dedicated and prepared. They conduct extensive research on their targets, identify their weaknesses, and plan their attacks. We know that members of Al Qaeda, other terrorist groups and the lone wolf extremist often receive training in the use or conduct of:

- Firearms, explosives and incendiary devices
- Biological agents, chemical substances, hazardous materials, and other contaminants
- Hand tools, small power tools, and other similar equipment
- False identities, disguises, and clandestine operations
- Reconnaissance, casing, surveillance, photography, and observation

Hate Groups, Militias, Cults, Radical Environmentalists and Opposition Movements

While some activist groups and cults are well-organized, most lack the organization, training, and funding that terrorist groups usually possess. As demonstrated by the recent protests in Egypt during the last year, the Internet has made it possible for even unaffiliated opposition movements to identify mutual targets of opportunity, coordinate their activities, plan joint or concurrent events, collaborate on common goals, and share information on tactics. Members of such groups sometimes receive training in:

- Firearms, explosives and incendiary devices
- Biological agents, chemical substances, hazardous materials, and other contaminants
- Non-violent protests and resisting arrest
- Sabotage and harassment

Vandals and Computer Hackers

Vandals normally do not have any special capabilities or training, and are likely to use whatever means and materials are readily available. Computer hackers, on the other hand, have extensive knowledge of IT systems, computer networks, system vulnerabilities, attack methodologies, malicious software, and IT security protocols. They have at their disposal and use a wide range of tools that assist them in probing systems, discovering exploitable weaknesses, launching attacks, and making it difficult for others to uncover and trace their actions. Hackers are also known to share technical and other information on system exploits over the internet.

Disgruntled Employees, Contractors, and Other Insiders

Insiders often possess one or more methods of accessing their targets (e.g., keys, combinations, passwords) that, when coupled with stolen confidential non-public information, can present a very real risk of undetected illegal activity within the asset. In many cases, employee and consultant insiders also have specialized technical or operating knowledge of relevant machinery, equipment, facilities, processes, procedures, access controls, security mechanisms and protocols, as well as countermeasures.

Mode of Attack – The Threat Spectrum

An attack against the energy sector could take one of the following forms:

- The physical destruction of an asset not associated with an explosive device
- The theft of property from the energy sector facility, parking lots, loading docks, etc.
- An Improvised Explosive Device (IED) or Vehicle-Borne Improvised Explosive Device (VBIED).
- Ballistic Attack with the goal of multiple casualties or damaging equipment.
- Vandalism of the facility or asset, to include defacement and damage of property.

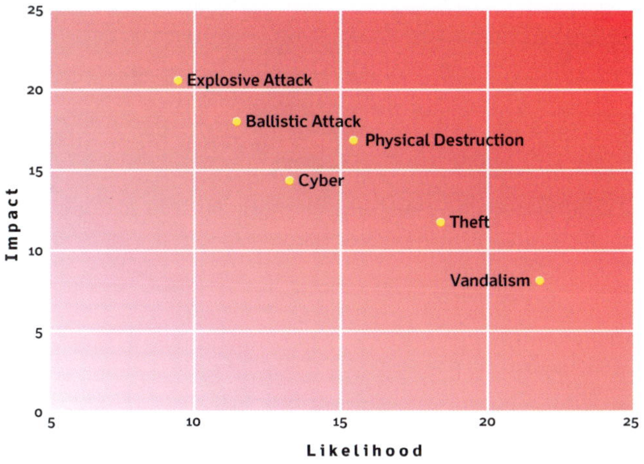

Potential Consequences

Based on the threat spectrum, the energy sector is subject to various consequences resulting from the mode of attack. The consequences include loss of system operation, destruction of assets, potential loss of life, etc. The potential for loss is heightened during economic and political turmoil, as well as seasonal peak demand weather conditions. Although these modes of attack are less likely than theft or vandalism, the impacts are high if successful.

Conclusion

The available information and recent experiences suggest that the energy sector critical infrastructure could be the target of a variety of adversaries. Terrorists and other adversaries seek out unprotected or "soft" targets; unfortunately, most of America's critical infrastructure currently qualifies as such. Facilities that fail to take proactive measures to evaluate their vulnerabilities, shore up their security systems, implement threat detection measures, update their emergency operations plans, and train their employees to handle incidents to reduce the likelihood of an attack and/or reduce the severity of its consequences.

Appendix III

Energy Sector CIP-014 Threat Assessment Template

The following CIP-014/R4 report is not complete as some of the sensitive information was redacted. As you can see, the table of contents contains a lot of information that would be provided in a finished report. I have kept as much information in the report so that the reader can use it as a template in constructing similar R4 reports

CONTENTS

Introduction

Infrastructure protection and security of the power grid is critical to our nation because electrical service is essential for public safety. As a result, the North American Electric Reliability Commission

(NERC) enacted the critical infrastructure protection (CIP) standard for physical security measures (CIP-014) to address physical security threats and vulnerabilities to the U.S. power grid and to reduce the overall susceptibility of physical attacks. The standard outlines a security approach that focuses on identifying the most critical facilities/assets and incorporates risk management planning to mitigate a range of threat profiles. The appropriate level and specific security measures employed by the infrastructure at each of its sites are unique; however, the basic principles of deter, detect/assess/communicate, delay, and response are valid for all locations based upon the principles of defense in depth.

Purpose and Objectives

Per NERC CIP 014, Requirement R4, the purpose of this report is to identify and document physical and operational weaknesses that could be exploited by an adversary(s) in order to compromise the infrastructure's ability to protect the bulk electric system that provides uninterrupted electrical services. The threat assessment defines the potential threats to the infrastructure's assets and is used as the roadmap for the vulnerability assessment. The vulnerability assessment (VA) identifies and evaluates those weaknesses across a broad range of threats/hazards and serves as the basis for developing the security master plan that outlines security upgrades, modifications of operational procedures, and/or policy changes designed to mitigate potential risks.

The results of this report will assist leaders in developing sound security decisions and corresponding investment justifications. A key element of these decisions is the determination of what constitutes an acceptable level of risk, since any risk or potential loss must be balanced against management expectations and the needs of our customers.

The benefits of conducting a threat and vulnerability assessment include:

- Provides current security posture of asset
- Highlights areas where greater security is needed
- Provides ability to prioritize improvements in a risk-based approach

- Assists in the justification of effective countermeasures
- Increases security awareness

The purpose of this plan is to evaluate, document, and recommend actions based upon the below listed objectives as stated in the NERC CIP 014 Reliability Standard:

- 4.1. Identify the unique characteristics of the identified and verified Transmission station(s), Transmission substation(s), and Primary control center(s).
- 4.2. Document and evaluate prior history of attacks on similar facilities taking into account the frequency, geographic proximity, and severity of past physical security related events.
- 4.3. Document and evaluate intelligence or threat warnings received from sources such as law enforcement, the Electric Reliability Organization (ERO), the Electricity Sector Information Sharing and Analysis Center (ES-ISAC), U.S. federal and/or Canadian governmental agencies, or their successors.

The objectives of the threat and vulnerability assessments are to determine whether current security countermeasures effectively address current, new and unforeseen threats. Based on the findings of this vulnerability assessment, the secondary objective of the assessment is to recommend enhancements to the current security posture, in order to ensure adequate protection of the asset(s).

In order to evaluate the potential threats and vulnerabilities of a physical attack, we followed a five-step process to ensure all elements of NERC CIP 014 were addressed.

Critical Assets

Per NERC CIP 014. Requirement R4.1, the evaluation shall consider the "unique characteristics" of the identified asset that, based on the threat assessment, would require protection from an adversarial attack. The vulnerability assessment answers the basic question, "what can go wrong should the system be exposed to threats and hazards of concern? Vulnerability assessments involve looking at key systems and determining how vulnerable they are to attack given a set of known and potential threats

During the R4 review and assessment the following assets were deemed critical to substation operations:

- Transformers
- PDC control shed
- Control block house
- Circuit breakers
- Communication tower
- Cable trenches
- CCVT/CPT transformers
- Communications vault

Likely Threats

A vital first step in conducting a vulnerability assessment for the energy sector's fixed assets is to perform a threat assessment. In performing a threat assessment, we assume that the asset is a target for attack. The term "target" means the asset has potential adversaries both inside and outside of its established boundaries. The term "attack" means any action intended to bring about harm, violence, damage or destruction. The primary goal of a threat assessment is to determine the likely threats against a given target based upon all available information and prior incidents. The information gathered to complete the threat assessment was utilized to complete the vulnerability assessment, the CARVER Matrix and Risk Reduction Recommendations.

History of Attacks

Per NERC CIP 014, Requirement R4.2, an evaluation of the "Prior History of Attacks" on similar facilities is required in order to identify and understand the physical security systems that should be designed to counter potential threats. To understand the attributes and capabilities of potential adversaries that could carry out malicious acts, it is necessary to look at the past history of incidents that have occurred, as well as gathering intelligence data from law enforcement or government agencies.

The following list is based upon those electric utility related security incidents that have occurred or reported in the U.S. since 2010:

Date/Location	History of Prior Attacks
2 April 2016, Tyngsborough, MA	FBI, with the assistance of the Massachusetts State Police, Chelmsford Police Department, Tyngsborough Police Department, and Army National Guard Civil Support Team, arrested Danny M. Kelly, age 61. Kelly was arrested following an investigation into incendiary devices found attached to power lines in Tyngsborough on 30 March 2016.
April 8, 2015, Newman Substation	The L3 fiber optic communication line was shot for the third time in a 24 month period. It took out communication from Newman substation to Afton substation for a short period of time.
January 2, 2015 (Report Date) Macho Springs Substation, Macho Springs, NM	Copper Theft - The chain link perimeter fence was cut from top to bottom to allow access into the substation. Multiple copper wires were cut and removed from the premises.
January 1, 2015 Luna Substation, Deming, NM	The south side perimeter fence was cut at the Luna Substation. The cut was about 5.5 foot high. The unknown individual (s) cut and removed the copper ground wire from equipment.
August 28, 2014 San Jose, CA	Thieves cut through a fence and stole equipment. Unlike in 2013, this second incident at the same PG&E transmission substation did not seem intended to disable the facility.

Date/Location	History of Prior Attacks
June 11, 2014 Nogales, AZ	Bomb planted at substation. Makeshift bomb left under a diesel tank at transmission substation owned by Unisource Energy Services. Bomb did not explode but did ignite a small fire. Service was not affected.
March 20, 2014, Sept 26, 2014 Ascarate Substation, El Paso, TX	The first field investigation was conducted on the incident that was reported on March 20, 2014 at approximately 9:44 am, by EPE personnel. Fence was cut and 4 marine batteries were reported missing from control room. The second field investigation was conducted on an incident that was reported on September 26, 2014 at approximately 1:00 pm. Fence was cut and 4 marine batteries were reported missing from control room.
February, 2014 Watertown, SD	Motor-operated switch shot at substation. East River's Ortley substation motor-operated switch was shot 15 times. No service outages or interruptions; substation sustained $7,000 in damages.
October 6, 2013 Lonoke County, Jacksonville, AR	Two power poles intentionally cut. Two utility poles were cut down using a stolen Sky Trim tractor. The power lines targeted in the attacks link a high-voltage transmission line with a switching station or substation. The incident resulted in a power outage.
October, 2013 Iroquois County, IL	Vandal damage: Eastern Illinois Electric Cooperative substation vandalized. Person(s) damaged the exterior and interior of a building.
September 29, 2013, Keo, AR	Fire at Keo substation. Entergy Arkansas' control house was destroyed by arson, no injuries or power outages resulted. Vandals' message at site read, "YOU SHOULD HAVE EXPECTED US". A suspect was apprehended and found responsible for the arson.
August 26, 2013 Lake Placid, FL	Fire damage. Glades Electrical Cooperative's Highlands Park substation severely damaged due to fire likely caused by gunshots. 1,300 customers lost power.
August, 2013 Cabot, AR	Bolts were removed from the base of a transmission tower, but the effort failed to bring down the tower.

Date/Location	History of Prior Attacks
August 21, 2013 Cabot, AR	A cable was attached to the framework of a transmission tower and across a railroad track. Perpetrators intended to use a passing train to bring down a transmission tower.
April 16, 2013 San Jose, CA	Attack on PG&E Metcalf substation. Phone lines to transmission substation were cut. 17 large transformers damaged by rifle fire. No service interruption or outages occurred. Substation sustained $15.4 million in damages and was disabled for almost a month. FBI investigation is ongoing.
December 16, 2012 Heber, AZ	Transformer shot at substation. Large power transformer at Navopache Electric Cooperative substation shot multiple times, causing a 10-hour blackout in the Heber and Overgaard communities and surrounding areas.
October 11, 2012, EPE Substation Chaparral, NM	Neutral copper cables cut and removed from all regulators on feeders 10, 11, and 12.
September 7, 2012 Tahlequah, OK	Substation shot by vandals. Four shots from high-powered rifle caused $1 million in damage to Lake Region Electric Cooperative substation. Two shots hit main transformer and two hit electric circuit breakers resulting in 2,000 customers losing power for less than 2 hours.
August 20, 2012, Montwood Substation, El Paso, TX	At approximately 5:00 AM a bullet hole was found in the voltage regulator, outer radiator fin at the Montwood Substation.
January 31, 2012 Little River, SC	Vandalism of substation. 4,500 customers lost service for six hours after attempted break-in at the Horry County Electric Cooperative substation. Specific details not released due to criminal investigation.

Date/Location	History of Prior Attacks
December 16, 2011, Active Shooter Incident at Southern California Edison	A gunman opened fire inside a Southern California Edison office building in Irwindale, California. The gunman, Andre Turner, killed two employees and wounded two others before committing suicide.
July 5, 2010 Hamilton County, IL	Intruder(s) drained oil from a large transformer. Alarms at dispatch center alerted workers who remotely opened circuit breakers.
March 31, 2011 Gillette, WY	Substation transformer shot at Powder River Entergy Corp. The vandalism caused more than 270 customers to lose power.

Evaluation of Potential Threats - NERC CIP 014, Requirement R4.2 & R4.3 require a threat assessment based upon a history of prior attacks and intelligence threat warnings.

Threat Spectrum

The first step was to develop a Threat Spectrum based on the impact of the below listed attack scenarios and their likelihood of happening. All possible situations could not be listed; however, the following list illustrates those scenarios most likely to occur:

- Explosives
- Ballistic attack
- Physical attack (sabotage)
- Cyber
- Vandalism
- Theft

Following is a visual depiction of the threat spectrum and as you can see vandalism and theft are the most likely adversarial events to occur at a substation. However, due to the minimal impact of these crimes, security recommendations were based upon higher consequence threats which can cause loss of life/injuries and or loss of system operations. A more detailed description of these threats is located in the Design Basis Threat section at the end of this chapter.

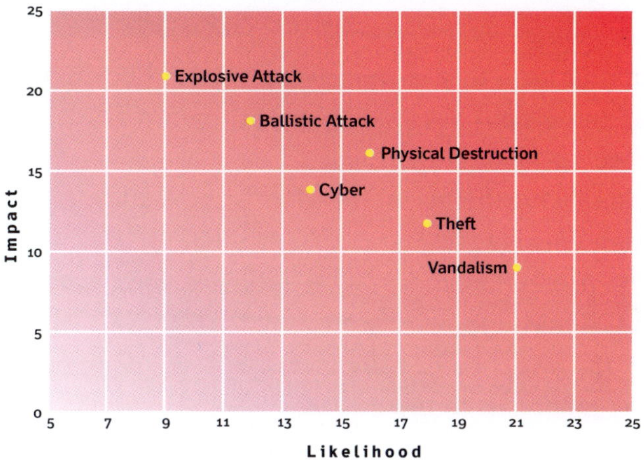

Design Basis Threats (DBTs)

The design basis threat (DBT) identifies capabilities that may include weapons, explosives, tools, and tactics that are useful in determining the detection, delay, and response criteria for the design and evaluation of an effective physical protection system. In the absence of a sufficiently detailed description of the threat it is difficult to determine with certainty the amount of protection that would be required to protect a given asset from the unacceptable consequences of an attacker.

Ballistic Attack: A ballistic attack by a determined adversary possessing the following attributes or capabilities:

- Possesses specialized shooting skills (possibly prior military) with sufficient knowledge to identify specific critical equipment, either from prior employment, observation, or open source research
- Suitable weapons, high powered rifle or assault rifle having effective long-range accuracy (7.62MM or 5.56 MM Caliber)
- Tools of entry or for otherwise destroying container integrity or features of the safe-guards system
- Vehicles which could be used for transporting personnel and equipment, as well as rapid retreat from the area prior to law enforcement arrival
- Intentionally fires multiple rounds at several targets with the goal of causing a large scale power outage

Vandalism: damage that is caused intentionally or unintentionally by an individual possessing the following attributes or capabilities:

- Lacking knowledge of the electric substation equipment to cause lasting damage
- May possess weapons with effective long-range accuracy (7.62MM Caliber)
- May throw debris (usually metal) into the equipment causing electrical components to trip
- May damage equipment/break in to steal something of value

Metal Theft by an individual or individuals to steal copper or other materials possessing the following attributes or capabilities:

- Lacking knowledge of the electric substation equipment to cause lasting damage
- May possess specialized tools of entry and/or used to cut and remove copper grounding wires

Insider Threat (disgruntled employee, contractor or vendor) possessing the following attributes or capabilities:

- Unpredictable employee that may be motivated for a number of reasons—a grievance against an electric utility based on a perceived injustice, passed over for promotion, disciplinary review etc.)
- May or may not possess advanced knowledge of electric operations; however, knows enough about the system to cause grave harm

Lone Wolf possessing the following attributes or capabilities:

- Unpredictable individual that may be motivated for a number of reasons—a grievance against an electric utility based on a perceived injustice, an infringement on personal property, a copycat type of attack to provoke or gain self-satisfaction or notoriety, or some cause unknown to anyone except the individual
- May or may not possess knowledge of electric operations; however, is motivated to research open source information on the internet

Activist possessing the following attributes or capabilities:

- Activists" have for years singled out the electricity sector as a target for protests, sabotage and vandalism, especially utilities owning nuclear or fossil assets
- Alleged grievances are usually directed at fossil fuel generating stations, transmission rights of way, water intake structures, nuclear plants, nuclear fuel storage, and upgrades in capacity to existing power lines (transmission and distribution)
- In many cases, these groups cover themselves with an "environmental blanket" to justify their illegal activities

Arson by an individual or individuals possessing the following attributes or capabilities:

- Unpredictable individual that may be motivated for a number of reasons—a grievance against an electric utility based on a perceived injustice, an infringement on personal property, a copycat type of attack to provoke or gain self-satisfaction or notoriety, or some cause unknown to anyone except the individual
- May or may not possess knowledge of electric operations

Drone Attack by an individual or individuals possessing the following attributes or capabilities:

- Unpredictable individual that may be motivated for a number of reasons—a grievance against an electric utility based on a perceived injustice, an infringement on personal property, a copycat type of attack to provoke or gain self-satisfaction or notoriety, or some cause unknown to anyone except the individual.
- May or may not possess knowledge of electric operations

Explosives (VBIED) by an individual or individuals possessing the following attributes or capabilities:

- Unpredictable individual that may be motivated for a number of reasons—a grievance against an electric utility based on a perceived injustice, an infringement on personal property, a copycat type of attack to provoke or gain self-satisfaction or notoriety, or some cause unknown to anyone except the individual.
- May or may not possess knowledge of electric operations
- Will have knowledge of explosives or incendiaries

Explosives (IED/Suspicious package) left near critical equipment by an individual or individuals possessing the following attributes or capabilities:

- Unpredictable individual that may be motivated for a number of reasons—a grievance against an electric utility based on a perceived injustice, an infringement on personal property, a copycat type of attack to provoke or gain self-satisfaction or notoriety, or some cause unknown to anyone except the individual.
- May or may not possess knowledge of electric operations
- Will have knowledge of explosives or incendiaries

CARVER Methodology

CARVER is an offensive target assessment tool originally developed for use by the U.S. military and the Intelligence Community in evaluating the vulnerabilities of enemy assets, and determining how best to exploit those vulnerabilities to attack a target.

The CARVER methodology addresses both physical security of a facility and the operational capabilities of terrorists or other adversaries by quantifying the likelihood of an attack based on target weaknesses. It acts as a target selection tool that uses a numerical ranking methodology to identify targets most attractive to attack by an adversary. To assess the likelihood of attack for a potential target, CARVER analyzes the following targeting criteria:

- **Criticality** – Single points of failure, Degree of importance to the system

- **Accessibility** – Ease of access to critical points of a facility or asset, and their exposure to adversarial acts
- **Recoverability** – The time and effort to recover system operation after an adverse event
- **Vulnerability** – Level of exposure to attack based on adversary capability
- **Effect** – Scope and magnitude of adverse consequences that would result from malicious actions and response to them
- **Recognizability** – Ability of an adversary to recognize assets, security systems, vulnerabilities, etc.

We completed the assessment through the eyes of an adversary, encompassing the following basic principles:

We examined all of the security programs from the terrorist's or adversary's viewpoint, (i.e., how would I, if I were the adversary, attack this asset and what would I target?)

We then determined what countermeasures would be required to prevent the terrorist or adversary from anticipating success and therefore launching such an operation.

We also assumed that conspicuous security procedures and equipment would not stop a determined terrorist or other adversary.

The relative risk for each threat and target combination is determined using algorithms to analyze potential consequences, relative importance of targets to the aggressor, and security vulnerability levels. Countermeasures are then developed to minimize the risk and; subsequently, the methodology is re-applied to determine if risk reduction is achieved. This iterative process is continued until the most cost-effective method of reducing the risk to an acceptable level is identified.

CARVER Matrix
(not provided as part of this redacted report)

CARVER Notional Rating Scale (Metrics)

The values 1-5 are used to calculate Risk and Probability of Attack (Pa) in the CARVER Matrix. The higher the number the higher both the Risk and Probability of Attack.

	C	A	R	V	E	R
5	Damage or loss of this asset will significantly impact/stop the asset's output or operations	Easily accessible, unlimited access with no security measures to prevent or detect	Extremely difficult to replace; long down time	A dedicated adversary with little capability and expertise can easily damage or destroy asset	Very high: Damage will have significant sociological, economic or reputational impacts, including loss of life and injuries	Easily recognizable and requires little or no training to identify
4	Damage or loss of this asset will greatly reduce the asset's output or operations	Accessible, limited access with some basic security measures in place to prevent entry (fence, doors, locks, etc.)	Difficult to replace with long down time	A dedicated adversary with some capability and expertise most likely will damage or destroy asset	High: Damage will have sociological, economic or reputational impacts, including some loss of life and injuries	Easily recognizable and requires a small amount of training to identify
3	Damage or loss of this asset will reduce the asset's output or operations	Somewhat accessible but the threat of observation exists. Some security measures are in place (e.g., CCTV or onsite personnel)	Replaceable with a relatively short downtime	A dedicated adversary with capability and expertise can easily damage or destroy asset	Moderate: Damage might have sociological, economic or reputational impacts with no loss of life but the possibility of injuries	Difficult to recognize and might be confused with other equipment and requires some training to identify
2	Damage or loss of this asset might reduce the asset's output or operations	Difficult to gain access; physical security barriers, observation and alarms are in place. Access granted to authorized personnel only	Easily replaceable in a brief time	A dedicated adversary most likely does not have the capability to damage or destroy asset	Low: Damage might have sociological, economic or reputational impacts with no loss of life or injuries	Difficult to recognize, easily confused with other equipment and requires extensive training to identify
1	Damage or loss of this asset will not impact output or operations	Not accessible; defined means of intervention in place	Immediate replacement; spare parts are immediately available or asset is redundant	A dedicated adversary does not have the capability and expertise to damage or destroy asset	None: Damage or loss of this asset will have no unfavorable impact	Unrecognizable except by experts

Criticality: A target is determined critical when an attack on the asset would result in significant impact on the output, mission, or operations. Metrics used for this analysis are:

Criticality Criteria	Scale
Damage or loss of this asset will have a significant impact/ stop the asset's output, mission, and/or ability to conduct operations.	5
Damage or loss of this asset will greatly reduce the asset's output, mission, and/or ability to conduct operations.	4
Damage or loss of this asset will reduce the asset's output, mission, and/or ability to conduct operations.	3
Damage or loss of this asset may reduce the asset's output, mission, and/or ability to conduct operations.	2
Damage or loss of this asset will not impact the asset's output, mission, and/or ability to conduct operations.	1

Accessibility: A target is accessible when an attacker can reach the target to conduct the attack and egress the target undetected. Accessibility is the openness of the target to the threat, or the ease in which the asset can be reached. Metrics used for this analysis are:

Accessibility Criteria	Scale
Easily Accessible: Attacker has relatively unlimited access to the target. (e.g., target is located outdoors with no effective security measures).	5
Accessible: Attacker has access to the target for an hour or less. Only limited specific information is available on the facility and the target. (e.g. target is located outdoor within a fenced perimeter or locked building, with no other effective security measures).	4
Partially or Somewhat Accessible: Attacker has access to the target but is under constant possible human observation. Some physical barriers and security measures may be present.	3
Hard/Difficult to Access: Human observation and physical barriers with an established means of effective security detection. Access generally restricted to operators or authorized persons. (e.g., inside building in a secured part of facility).	2
Not Accessible/Very Difficult: Physical barriers, alarms, and human observation. Defined means of intervention in place.	1

Recoverability: A target's recoverability is measured in the time it will take for the specific system to be repaired, replaced, or recover. Metrics used for this analysis are

Recoverability Criteria	Scale
Extremely difficult to replace long down time > 1 year	5
Difficult to replace with long down time 1 months to 1 year	4
Can be replace in a relatively short time 1 month to 1 week	3
Easily replaced 1 week to 1 day	2
Immediate replacement. Spare parts are readily available or asset redundancy < 1 day	1

Vulnerability: A target is vulnerable when an adversary can damage, destroy, compromise, or exploit the asset using available capabilities. Metrics used for this analysis are:

Vulnerability Criteria	Scale
A dedicated adversary with little capability and expertise can easily damage or destroy the asset.	5
A dedicated adversary with some capability and expertise will most likely damage or destroy the asset.	4
A dedicated adversary with capability and expertise to attack will damage or destroy the asset.	3
A dedicated adversary most likely does not have the capability or expertise to damage or destroy the asset.	2
A dedicated adversary does not have the capability or expertise to damage or destroy the asset.	1

Effect: Effect is a measure of the scope and magnitude of an attack or the adverse consequences that would result if the asset were destroyed. Metrics used for this analysis are:

Effect Criteria	Scale
Very High: Damage or loss of this asset will have significant sociological, economic, political, or reputational impacts, including loss of life and injuries	5
High: Damage or loss of this asset will have sociological, economic, political, or reputational impacts, including some loss of life and injuries.	4
Moderate: Damage or loss of this asset may have sociological, economic, political, or reputational impacts, with no loss of life but some injuries may occur.	3
Low: Damage or loss of this asset will have a significant sociological, economic, political, or reputational risks, with no loss of life or injuries	2
None: Damage or loss of this asset will have no unfavorable impact.	1

Recognizability: A target's recognizability is the degree to which it can be identified by an attacker as a valuable target without confusion. Metrics used for this analysis are:

Recognizability Criteria	Scale
The target is clearly recognizable and requires little or no training for recognition.	5
The target is easily recognizable and requires only a small amount of training for recognition.	4
The target is difficult to recognize or might be confused with other targets or target components and requires some training for recognition.	3
The target is difficult to recognize. It is easily confused with other targets or components and requires extensive training for recognition.	2
The target cannot be recognized under any conditions, except by experts	1

Appendix IV

Leo Labaj Biography

When I was a young boy, around seven or eight years old, I asked my mother a question that had been bothering me for some time: Why was my last name different from hers? I could see immediately that my question distressed her – and from my current perspective I realize how my mother must have dreaded the moment she knew would arise sooner or later. But she did respond, and her answer changed my life forever. My mother explained that my father, Leo Labaj, Sr., (pictured on right) had been killed in World War II, back when I was only 18 months old. She remarried after the war and took her new husband's last name.

I learned that my father had been serving in the U.S. Navy and was stationed aboard the USS Morris, DD-417, the flagship for Destroyer Squadron 2 (DESRON2) in the Pacific Theater. The Morris earned 15 battle stars during the war. On March 6, 1945, while patrolling station A-11 off the island of Okinawa, a Japanese "Kate" (Kamikaze) aircraft, carrying either a heavy bomb or torpedo, closed in on the Morris. The ship's antiaircraft guns] scored hits and set the plane afire, but those hits failed to stop the suicidal pilot from pressing his attack. Shortly after 1815 hours, the Kate crashed into the ship on the port side between the No. 1 and No. 2 guns. The explosions caused by the impact ignited fires, which spread quickly. The surviving crew needed two hours to bring the fires under control and another 30 minutes to extinguish them. One-fourth of the Morris's crew was killed in the attack,

USS *Morris*

including my father, stationed inside No. 1 gun. He died instantly when the Kamikaze hit the ship.

My mother gave me the memorabilia the Navy had sent her, which included my father's service medals, a large picture of the Morris and the shellback certificate he had earned crossing the equator on the ship, as well as some of his personal effects. During those discussions with my mother, I felt overwhelmingly proud of my biological father, though I also loved my stepfather.

I have attended several USS Morris reunions and had the opportunity to meet some of the sailors who knew my father. Sad to

say, they're all gone now. During the last 10 years, I have spoken with many of the ship's sailors, and they all told me Leo, Sr., was very well liked by everyone who served with him. My father was the ship's coxswain and thus piloted the liberty boat whenever the Morris docked at a port. As you might imagine, this made him popular among the crew. The photo at left, taken at a Morris reunion in 2013 in Norfolk, Virginia, shows three of the ship's sailors extinguishing candles for their fellow crewmembers who were lost. As they did, they called out each sailor's name. It was especially moving for me when they put out the candle for my father. The reunions turned out to be a godsend, because I got to know something of my father through his friends and shipmates.

A Navy Career

Perhaps as a tribute to my father's memory, I eventually decided to join the Navy. My mother, encouraged by my stepfather, wanted me to attend the Naval Academy after I graduated from high school. She thought my father's death in the line of duty would help my case for admission. But I threw a monkey wrench into the process when I quit school before graduating and enlisted in the Navy.

The truth is I had never liked school, finding it constantly boring. So, when my two best friends Walter Lamory and Frank Camarotti – who were a couple of years older than I and had already graduated high school – decided to join the Navy, I dropped out and went with them. The year was 1961. The Cold War with the Soviet Union was in full swing, and the Vietnam War was just beginning.

Attempting to boost enlistments at the time, the Navy employed a recruitment strategy called the "Buddy Program." The idea was to encourage groups of young men to join the service together. In exchange, the Navy agreed to assign the buddies to the same company at boot camp. That they did, and Walter, Frank and I remained together until graduation, when the Navy sent us off to different technical schools. (This photo was taken at boot camp graduation, I on the right, Frank in center and Walter on the left.) I learned something important about myself during that time. I noticed that while most of my fellow boot campers were already counting the days until their enlistments were over, I was loving every minute of it. I was even nominated as the company's Mastered-at-Arms in my first days at boot camp and Honor Man at graduation. I knew I had found a home in the Navy.

Following a brief home leave, we went our separate ways to our new assignments. I ended up in Aviation Ordnance "A" School in Jacksonville, Florida, while Walter studied to be a Fire Control Technician at the Great Lakes Training Center near Chicago, and Frank went to Radar Technician A School, also at the Great Lakes center.

For the record, I came to regret quitting school, but I never regretted joining the Navy.

Sea Duty

I spent the next eight years as an aviation ordnanceman (AO), in other words a bomb loader or, as affectionately known, a BB stacker for Navy fighter aircraft. Following my initial training I was assigned to Fighter Squadron 31 (VF-31) at Naval Air Station (NAS) Cecil Field, Florida. My very first deployment as a member of the "Tomcatters," our squadron

nickname, was servicing F-3 Demons aboard the carrier USS Essex. The Essex was my first deployment aboard a naval ship. I had just come out of boot camp and went to my initial training at Naval Air Station, Key West, Florida. Following a short training stint at Key

West, I deployed aboard the USS Essex for what was called carrier qualifications. Some background on the Essex, it was a wooden decked aircraft carrier and the lead ship of the 24-ship Essex class built for the United States Navy during World War II. She was the fourth US Navy ship to bear the name. Commissioned in December 1942, Essex participated in several campaigns in the Pacific Theater of Operations, earning the Presidential Unit Citation

and 13 battle stars. Decommissioned shortly after the end of the war, she was modernized and recommissioned in the early 1950s as an attack carrier (CVA), and then eventually became an antisubmarine aircraft carrier (CVS). In her second career, she served mainly in the Atlantic, playing a role in the Cuban Missile Crisis. She also participated in the Korean War, earning four battle stars and the Navy Unit Commendation. She was the primary recovery carrier for the Apollo 7 space mission. A humorous although embarrassing event occurred while on this very short deployment. During our training the ship went to General Quarters (battle stations) and all were supposed to report to their duty stations. As soon as the alarm sounded I made my way to the flight deck from below decks and ran into a crusty old boatswain mate. He asked me where I was going and I told him I was going upstairs when I should've said I was going topside. Using incorrect jargon on a Navy ship is a no-no and I took a lot of heat from that episode throughout the remainder of the cruise.

In October 1962, we were assigned to the Cuban blockade aboard the USS Saratoga. Earlier that year, when our spy planes had photographed Russian missiles and their launchers in Cuba,

President John F. Kennedy ordered the Navy to prevent ships from the Soviet Union and its allies from reaching the island country until the missiles were withdrawn. The standoff lasted 13 days. Conducting the blockade was the closest we young sailors had come to a war footing. Our aircraft were loaded with live missiles and other ordnance for possible use against the ships and planes of Cuba and the Soviet Union. At the end of the blockade, the Russians stood down and their ships left the area. The Saratoga sailed into Guantánamo Bay prior to returning to Jacksonville.

By that time, I had become so enamored of the Navy that I reenlisted during our stay at Gitmo, volunteering for another six years.

More Sea Duty

Following my reenlistment, and a home leave, I returned to Jacksonville to attend Aviation Ordnance "B" School. After completion, I was promoted to Aviation Ordnanceman 2nd Class and was assigned to training squadron VF-121 at Miramar Naval Air Station near San Diego. This training involved transitioning from the F-3 Demon to the F-4 Phantom fighter aircraft. In December 1963, I received my first F-4 operational assignment in VFA-213, aka the Black Lions, also stationed at Miramar. In February 1964, VFA 213 accepted the first of the new F-4B Phantom IIs. In 1965, VFA-213 joined Carrier Air Wing 11 (CVW-11) assigned to the USS Kitty Hawk, where I served from October 1965 to June 1966 in my first deployment to the Vietnam Theater.

During that deployment, and shortly after leaving Pearl Harbor, Hawaii, bound for the Western Pacific, the Kitty Hawk suffered an on-board fire. It was located in the mess deck and became a news story throughout the United States. I was married by then, and my wife was staying with my mother. They both learned about the fire on TV. While they were watching the news coverage, the doorbell rang. It was Western Union delivering a telegram. My wife wrote me later that my mother almost fainted, because she feared the telegram was about me – just as she had received a telegram informing her that her husband, my father, was killed aboard the USS Morris. It turned out the telegram was only to inform my wife that she had failed to notify the lienholder of our new car that she had moved out of state.

My second deployment to the Vietnam Theater was aboard the USS America attached to VF 33. This would turn out to be a seven-month cruise. On April 10, 1968, the America sailed out of Norfolk bound for "Yankee Station," off the coast of Vietnam. The next day, the ship completed its complement of men and equipment when the remainder of Air Wing CVW-6's aircraft joined us at sea. And on May 31, the America arrived at Yankee Station, where we began combat operations. During four line periods – about 50 days each – on Yankee Station, the America's aircraft attacked various targets, attempting to impede the flow of men and war materiel to the south.

On July 10, pilot Lt. Roy Cash, Jr. (Johnny Cash's Nephew), and radar intercept officer Lt. J.G. Joseph E. Kain, Jr., in an F-4J Phantom from VF-33, downed a MiG-21 "Fishbed" fighter some 17 miles northwest of Vinh, North Vietnam. It marked the ship's first MiG kill. The America and Air Wing CVW-6 would later be awarded the Navy Unit Commendation for their performance during that time.

Related to the downing of the MiG-21, I was a senior quality-control petty officer for VF-33. My responsibility was to ensure that all ordnance was functional and loaded correctly on strike aircraft. Prior to each aircraft taxiing to the catapult, I function-checked its wing-loaded Sidewinder missiles with a flashlight to ensure they could detect a heat source. In the case of Cash's and Kain's F-4J, I discovered that the No. 1 missile, located on the left side of the aircraft, did not pass the check. Although the catapult officer was pressuring the load team to get the aircraft to the catapult, I halted the Phantom on the deck until the missile was changed. The switch took only a few minutes, but it turned out to be critical

to the success of the mission. Cash and Kain shot the MiG down using the very missile I had ordered changed on the flight deck.

Between line periods, the America visited Hong Kong, Yokosuka and Subic Bay. With her mission on Yankee Station nearing completion, we launched the last of our attack missions on October 28. The next day, we set sail for Subic Bay to offload various Yankee Station assets. At the time, my enlistment was nearing its end as well. I had planned to get out of the Navy and return to civilian life. So, instead of staying on the ship all the way back to Norfolk, I elected to catch a Navy aircraft from Subic Bay to San Francisco. From there it was on to Boston Naval Station where I received my discharge.

Looking back, I consider my Navy career a success. It gave me the underpinnings to succeed in life, and it provided some of the key components I would need for my next career, in the CIA. Shortly after my discharge, I received a commendation from the Commander in Chief of the United States Pacific Fleet.

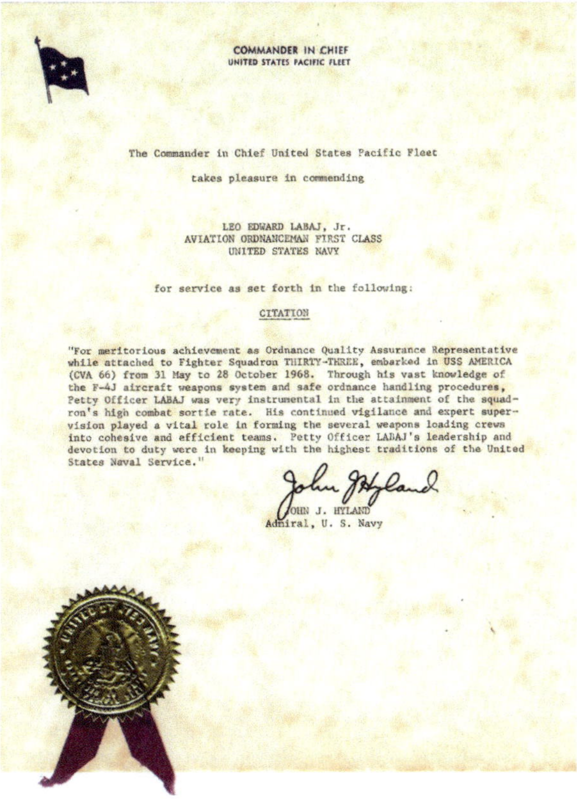

COMMANDER IN CHIEF
UNITED STATES PACIFIC FLEET

The Commander in Chief United States Pacific Fleet

takes pleasure in commending

LEO EDWARD LABAJ, Jr.
AVIATION ORDNANCEMAN FIRST CLASS
UNITED STATES NAVY

for service as set forth in the following:

CITATION

"For meritorious achievement as Ordnance Quality Assurance Representative while attached to Fighter Squadron THIRTY-THREE, embarked in USS AMERICA (CVA 66) from 31 May to 28 October 1968. Through his vast knowledge of the F-4J aircraft weapons system and safe ordnance handling procedures, Petty Officer LABAJ was very instrumental in the attainment of the squadron's high combat sortie rate. His continued vigilance and expert supervision played a vital role in forming the several weapons loading crews into cohesive and efficient teams. Petty Officer LABAJ's leadership and devotion to duty were in keeping with the highest traditions of the United States Naval Service."

John J. Hyland
JOHN J. HYLAND
Admiral, U. S. Navy

Civilian Life

After my discharge, I returned to my hometown, Chicopee, Massachusetts, where I immediately began looking for a civilian job. By then I had a wife and two children to support.

I quickly found out, however, that civilian life wasn't very satisfying for me. For a few months, I worked at Colt's Manufacturing Company in Hartford, Connecticut, helping to produce M16 rifles destined for our troops in Vietnam. Essentially, I was repeating what I did during my naval service: quality control. At the time, Colt was operating at a feverish pace, 24 hours a day/7 days a week, to turn out the M16s, and quality control was a big part of that process. The problem for me was that the West Hartford plant was about 70 miles from Chicopee, so the daily commute was grueling. I soon decided I needed something closer and, maybe, a little more exciting. As an interim step, I joined the Connecticut Air National Guard at Bradley Field, where I had intended to do my one weekend a month and two weeks of active duty each summer. But a fantastic opportunity suddenly presented itself. During my interview with an officer in the munitions section, I was told they needed an additional bomb-disposal technician. If I was interested, he said, they would send me to Explosive Ordnance Disposal School. I readily accepted. After five months as a civilian, I would be returning to a Navy school, though as a U.S. Air Force Technical Sergeant (E-6), studying at one of the service's toughest schools, where the dropout/failure rate was approximately 40 percent.

Located in Indianhead Maryland, EODS was the boot camp for bomb disposal. After graduation, EOD technicians are required to be proficient in detecting, identifying, rendering safe, recovering and performing field evaluations of any explosive ordnance. That includes ordnance that has been fired, damaged, dropped or placed in such a position as to be hazardous to installations, personnel or material. Explosive ordnance includes all munitions, both domestic and foreign, as well as improvised, chemical, biological, nuclear and incendiary. Despite the challenges, I completed the training at the top of my class.

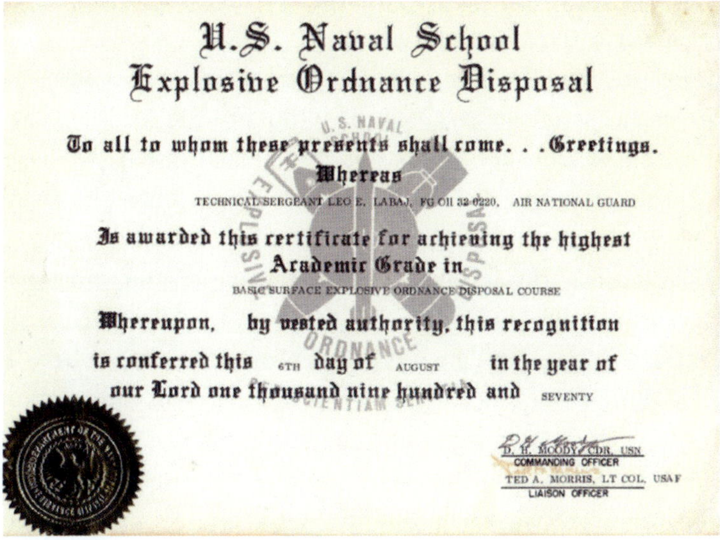

U.S. Naval School
Explosive Ordnance Disposal

To all to whom these presents shall come. . .Greetings.

Whereas

TECHNICAL SERGEANT LEO E. LABAJ, FG 011 32-0220, AIR NATIONAL GUARD

Is awarded this certificate for achieving the highest Academic Grade in

BASIC SURFACE EXPLOSIVE ORDNANCE DISPOSAL COURSE

Whereupon, by vested authority, this recognition is conferred this 6TH day of AUGUST in the year of our Lord one thousand nine hundred and SEVENTY

D. H. MOODY, CDR, USN
COMMANDING OFFICER

TED A. MORRIS, LT COL, USAF
LIAISON OFFICER

After six months at EODS, I returned to the Connecticut Air Guard as an Air Technician, a full-time job that increased the Guard's explosive ordnance disposal (EOD) capability. The Guard's operational mission required alert aircraft to be armed at all times whether on the tarmac or in a hanger, thus the need for EOD capability. We were also required to respond to improvised explosive device (IEDs) incidents in support of law enforcement and other military units in Western Connecticut.

Not long after returning to Bradley Field, I experienced another life-changing event when the Guard lost its Air Defense mission. Suddenly I held substantial training in a field that I had grown to love but no longer had an opportunity to use. So, I started shopping my services around. When I learned that the CIA was looking for what they called "Bomb Techs," I contacted them. I received the paperwork to apply, but to my dismay I learned it would take a year to complete the process. During that time, while constantly watching the mail and waiting for my acceptance, I continued in the Air Guard, though not in an EOD position. I went back to serving as a bomb loader on fighter aircraft. In January 1972 I was accepted into the CIA to begin my new career.

More Career Moves in the Offing

My career in the CIA spanned 22 years. During that time, I had three overseas assignments, including tours in the Middle East, Latin America and Europe. The assignments each lasted two years or more. I also served in 76 countries on a more limited basis. And I had the opportunity to conduct vulnerability assessments on high-threat residences, borders, critical infrastructure and more than a hundred international airports and seaports. I worked primarily on counterterrorism, except for my Latin American tour, where I performed counterinsurgency activities, primarily arms interdiction.

Another interesting aspect of my career in CIA was my "bomb tech" experience which provided me an opportunity to travel to many bombing incidents in which the US had an interest. As you can guess the 70s and 80s were very interesting years in my career.

When I retired from CIA in 1993, I went to work at the Y12 National Security Program Office in Oak Ridge, Tennessee. There I became involved in research and development of security related products. The Y12 highly secure facility provided storage for most of the weapons-grade enriched uranium in the United States. The security technologies I helped develop for the Department of Energy would be destined for use by Y12 and other sensitive DOE facilities nationwide. One major development was the Enclosed Space Detection System (ESDS), which could detect the heartbeat of human beings in vehicles to include fully loaded tractor-trailers without unloading their contents. I was the program manager for that technology, which is still being used extensively today at international border crossings and prisons. In 1997, my team and I received the R&D 100 award for the ESDS as one of the 100 most significant products developed that year.

I left Y12 in 1999 to start my own company, Advanced Security Products. The company was created to market, install, and train in the use of the ESDS, which became known as the Heartbeat Detector. The Heartbeat Detector was eventually licensed to Mark White, former governor of Texas, who started a company called GeoVox and marketed the product under the name Avian. GeoVox, under Gov. White, became successful in improving the heartbeat detector and selling it both nationally and internationally.

In the fall of 2001, not long after the 9/11 terror attacks, I was invited to join a company called Security Management International, which was acquired by Telemus, a company founded by Jon Monett, my former supervisor during my Agency career. I started work at Telemus as a consultant, and my first order of business was to conduct a water-vulnerability assessment for the Utility Department of Bellevue, Washington. The assessment went well; so well that it generated many more such projects in response to the Safe Drinking Water Act administered by the Environmental Protection Agency. In all, over the next 10 years, I conducted approximately 50 water-vulnerability assessments both nationally and internationally for Telemus.

Telemus SOLUTIONS

SECURITY THROUGH INTELLIGENCE

Although I started as a consultant, it wasn't long before I became a full-time employee of Telemus, which had changed its name to Telemus Solutions. I was promoted to Vice President for Infrastructure Protection Services and led a team that conducted hundreds of vulnerability assessments during my tenure there. Telemus Solutions was acquired by Raytheon in 2008, and the company became Raytheon Telemus. I was retained by Raytheon as Vice President of the Critical Infrastructure Protection Branch. I remained with the company until 2010, when I left to become a consultant for the new Security Management International. I continue there to this day, conducting training in target analysis and vulnerability assessments in both the United States and internationally.

Raytheon Telemus

Acknowledgments

During my stint in the Special Devices Branch and later Special Activities Division, I had the opportunity to be mentored by some of its legends, including Arthur "Mick" Donahue, who was awarded "Trailblazer" status in 2011 by then Director of Central Intelligence David H. Petraeus.

> *Throughout his lengthy career, Mick, who died in 2004, served as a Technical Operations Officer with the Office of Technical Service. He had an outsized impact on a broad range of complex intelligence requirements. Donahue spent 33 years conducting and leading technical operations in support of intelligence collection, special operations, paramilitary programs, counterintelligence, counterterrorism and counternarcotics operations. His remarkable achievements, which provided the inspiration for and innovation behind many of the most important and noteworthy technical ops of the late Cold War period – most of which remain classified – exemplify the qualities of a Trailblazer. Donahue's commitment to technical innovation, Intelligence Community engagement, robust target analysis, tradecraft and the conduct of complex counterterrorism and counterinsurgency operations, continues to pay dividends today.*

I mention Mick because I consider him the godfather of vulnerability assessments. He perpetually advocated target analysis and vulnerability assessments, and he created a mindset among all of us Techs.

I again want to acknowledge Jon Monett, my friend, colleague and former supervisor. To this day, I look upon Jon and on Mick Donahue as my mentors.

SMI

Printed in Great Britain
by Amazon

28547453R00110